SOVIET SPACE EXPLORATION
The First Decade

SOVIET SPACE EXPLORATION

The First Decade

by William Shelton

Introduction by
Cosmonaut Gherman Titov

ARTHUR BARKER LIMITED
5 Winsley Street London W1

Copyright © 1968 by William Shelton
First published in Great Britain in 1969

SBN 213 17799 4

Printed in Great Britain by
Lowe & Brydone (Printers) Ltd., London

Acknowledgments

A number of individuals of Novosti Press Agency in Moscow and of the Soviet Embassy in Washington were particularly helpful in providing extensive material on the Russian space program and in arranging for my travel to major science centers in the Soviet Union. I wish especially to thank Boris Romanov, Aleksei Belakon, Georgi Isachenko, and Boris Sedov of the Washington staff, and Yevgeny Ruzhnikov, who fulfilled numerous specific requests and served as escort in the U.S.S.R.

For background briefings on Soviet astronautics, I am indebted to the U.S. Air Force, the Library of Congress, and the Central Intelligence Agency, as well as to Dr. Charles Sheldon, formerly of the President's Aeronautics and Space Council, and to Dr. Fermin Kreiger and Thomas Wolfe of the Rand Corporation, Santa Monica, Calif.

The writer, Martin Caidin, was the first American journalist to recognize that the Soviet space program represented a major scientific thrust. His book *Red Star in Space* (Crowell-Collier Press, 1963) was of considerable help as the first authoritative interpretation of Soviet astronautics. I wish also to thank Martin Caidin for permission to use portions of his interpretations of *I Am Eagle*, by the second Soviet cosmonaut, Gherman Titov.

I would also like to thank my friend and neighbor Colonel John Glenn for his letter to Cosmonaut Titov suggesting that he write the introduction for this book. I am grateful to Cosmonaut Titov, whose writing ability is especially noteworthy, for his introduction.

Finally, I am indebted to Soviet space expert and interpreter Joseph Zygielbaum, president of Terra Space Corporation of Malibu, Calif., for the numerous translations of technical documents he furnished me and especially for the abundant suggestions and criticisms he made after reviewing the first-draft manuscript.

Contents

	Introduction	ix
1	The Raising of the Curtain	1
2	The Trailblazers	16
3	The Treasure Trove of War	34
4	Before Sputnik	45
5	October 4, 1957	52
6	The Head Start	64
7	The Young Man From Klushino	81
8	The Gagarin Flight	93
9	The Poet of Space	102
10	Vostok and Its Pilots	124
11	"I Am Seagull"	147
12	Flights of the Voskhods	160
13	Automatic Earth Satellites	190
14	To the Moon and Planets	213
15	Military Applications	238
16	Death in Space	265
17	Soyus	278
	Chronology of Soviet Space Flights	284
	Selected Bibliography	325
	Index	329

Introduction

We Shall Adhere to Our Principles

by GHERMAN TITOV
Pilot-Cosmonaut, Hero of the Soviet Union

The book, of course, contains debatable points. Some apparently result from inadequacy of information. Others, perhaps, stem from a desire to find too many analogies in the space programs of the U.S.S.R. and the U.S. But these points do not obscure the merits of the book.

The chief merit is that it acquaints the reader with the real, impelling reason for all Soviet space exploration and, if I may say so, with the "secret" of the Soviet achievements in space. The author shows that Soviet space explorers are moved by one aim: to gain maximum scientific information about space in general, and the planets of the solar system in particular, with the least expenditure of effort and resources. That is why our space research is conducted in such a way that every new experiment helps to solve some essentially new problem. Soviet scientists try not to duplicate solutions already attained. They concentrate efforts on promoting a next stage in fathoming the universe by technical devices or by man himself.

This attitude of the Soviet space explorers precludes their participation in any space exploration race. It much rather calls for cooperation of Soviet researchers with their colleagues in other countries. Hence the active participation of Soviet scientists and cosmonauts in international forums on problems of peaceful exploration of outer space. When Soviet scientists mention their priority in many space achievements, I think they are prompted by pride for their state which provides them with all the necessary means and facilities.

In the decade after the beep-beeps of the first Soviet Sputnik the principles adhered to by Soviet space explorers helped them to cross out many question marks which loomed before October 4, 1957. No one doubts now that man can live and work under weightlessness, both inside and outside a spaceship. It also has been proved that spacecraft can land so softly on other planets that the most delicate scientific instruments remain undisturbed. A way has been found toward overcoming the limits of the power resources for modern carrier-rockets. It is no longer necessary to build huge interplanetary ships right on the earth. With the help of automatic docking they can be assembled from smaller units orbiting the earth. No one today argues the ability of a cosmonaut to maintain permanent communications with the earth from other planets or from distant regions of outer space.

The remarkable experiments of the U.S. explorers confirm these and all other findings obtained for science by Soviet space investigators.

In the next decade space explorers of my country will continue to adhere to principles explained here. We shall center our efforts on the solution of key problems. Our scientists are working along this line. The same is attended to by designers of spacecraft and scientific equipment. We, members of the cosmonaut detachment, are also getting ready for that. I think that in the second decade of the space era the Soviet Union's achievements in space exploration will be just as impressive as those described in this book.

SOVIET SPACE EXPLORATION

The First Decade

I
The Raising of the Curtain

OCTOBER 4, 1957. A huge white rocket stands poised on a remote plain in southwestern Russia, a shiny 184-pound sphere, 23 inches in diameter, nestling in its nose. The prevailing wind bends and smooths desert grasses; a few field sparrows flit and chirp. The nearest town, Tyuratam, lies 80 miles east of the Aral Sea.

About 150 yards upwind, a group of men wait in a blockhouse. Their leader, a calm stocky man with a receding hairline and deep-set brown eyes, stands alone beside a periscope window, dressed in a rumpled gray suit. Sergei Pavlovich Korolev, the son of a Zhitomir schoolteacher and "chief constructor," hears the last of the imperious cadence of the countdown—*tri, dva, odin.*

A blast of flame erupts at the rocket's base, followed by the delayed crackling thunder of ignition. The three-stage, liquid-propelled rocket ascends ever faster on its pulsing tail of fire, climbs toward the upper atmosphere, tilts over to the horizontal on a heading toward China, and is finally lost to vision. Overhead, its condensation trail assumes the shape of a pretzel as a loose and murky cloud hangs over the ground at the blast-off point. Finally, the field sparrows begin to chirp again.

Sputnik 1, pretender to the starry throne of space, is on its way. Within 90 minutes, its "beep, beep" is heard round the world. In Western nations, dials spin frantically as huge instruments search for the alien cry of the first artificial satellite of earth. Its cry is confirmed. The sphere is registered. An event is occurring that is to change the course and nature of history. Russia has raised the curtain on the age of space.

A month later, on the far side of the world, a small group of us watched another great nation's attempt to launch the satellite-carrying U.S. Vanguard rocket known as TV–3 end in a humiliating and fiery pad explosion. To redeem the failure another month later, a new Vanguard fumed and smoked on its pad at old Cape Canaveral on one of the coldest January nights on the Florida record. After innumerable delays, the countdown was halted once again, at an agonizing 22 seconds before launch. During the frustrating hold, a newsman suddenly shouted, and those of us waiting and praying for Vanguard gazed upward to see—almost as a vision—the second Russian satellite, *Sputnik 2*, weighing 1,120 pounds, coursing overhead in the night sky like a cold and glowing celestial ghost. It held the Russian dog Laika, the first living creature to venture into orbit, now dead.

On July 9, 1958, nine months after the Soviet Union astounded the world with the successful orbit of *Sputnik 1*, Premier Nikita Khrushchev announced to a group assembled in East Berlin: "For those who created the rockets and artificial earth satellites, we will raise an obelisk and inscribe their glorious names on it in gold so they will be known to future generations in the centuries to come. . . . We value and respect these people highly and assure their security from enemy agents who might be sent to destroy these outstanding people, our valuable cadres." The promised commemorative, weighing 240 tons, now rises from Moscow's Mira Prospekt as a gracefully upswept fin of stone and titanium capped with a gleaming rocket.

The emotional impact of Russia's sudden successes in the new field of space science and technology, and the diverse implications of her success to the world as a whole, are discussed with heat and urgency with every new Soviet breakthrough. But while there is little doubt that the Soviets themselves could hardly be surprised by their own efforts, nevertheless their rate of success, and with it the rapidity with which they achieved renewed stature as a modern technological power, was as striking to the average Soviet as it was to the average American.

The true impact of the amazing rate of development of Soviet space science in the post–World War II years must be appraised partly within the strange ironies of her "victory" of World War II. It was a strange victory indeed. Despite the superb but traditional Soviet field equipment, her splendid automatic rifles, Tiger tanks, and Katyusha artillery rockets, the Soviets emerged at the moment of victory to see their expedient field ally and ideological enemy, the United States, even more demonstrably superior scientifically, with her early A–bomb monopoly. And even the defeated enemy, Germany, had just brandished a potent new technological wonder weapon, the powerful V–2s, which had rained down on London and Antwerp in the closing months of the war.

The Russian moment of victory was, in the eyes of the Soviets, not so much the moment of longed-for security as a bitter moment of proof that they were second-rate technologically. To them, it seemed inevitable that the German-developed rocket technology and U.S.-developed atomic explosives would be joined in a single, dominating wonder weapon, against which there was then no known defense. Indeed, the fact that they were not immediately joined and that there was an inordinate delay in rocket development in the U.S., was, for the Russians, the greatest—and most beneficial—surprise to follow the hollowest victory in their history.

It was thus the terror implied in the way World War II concluded that initially motivated the Russians to make a maximum and largely secret effort to catch up in rocket and nuclear technology. This impetus enabled them in the first decade after *Sputnik 1* to launch into space over 200 instrumented or manned vehicles whose total mass weight exceeded 2 million pounds (compared to about 1.4 million pounds for the U.S.). Their launch frequency for payloads of all types quadrupled between 1960 and 1962, doubled in 1964, and climbed to still higher records in 1966 and 1967.

Soviet space successes, especially the early ones, gave the Soviets what a succession of stringent Five-Year Plans, constant sacrifices, and even Olympic records could not give them—a new rallying point for long-disturbed national feelings—and it provided them with a firm new image of themselves to which they could respond patriotically, emotionally, and even, to a certain degree, spiritually. Internally, the Soviet response to the lure of the cosmos was powerful and fervent. Internationally, space achievements won them such immediately recognizable respect that their adept propagandists quickly formulated two new myths: space milestones, they claimed, had suddenly made them the world's leading technological nation; and achievements in space were, in their eyes at least, proof of a superior form of government. Although these were merely skillfully contrived myths, they were successfully implanted in the Orient, in South America, and in some of the emergent nations in Africa.

But it was the initial reaction in the United States that most pleased the Soviets, who had for so long endured the underdog role in science and technology. The early Sputniks caused far more consternation, soul-searching, and restive debate in all levels of American culture than even the most optimistic Russians could have predicted.

They were overjoyed when several members of Congress called *Sputnik 1* a "hoax," and initiated the growth of an incredible series of myths with which we have surrounded the Soviet space program. When tracking stations outside the Soviet Union finally confirmed the unexpected and disconcerting event of *Sputnik*, President Dwight Eisenhower innocently perpetrated the first myth on the American public. In answer to a reporter's question at a press conference, on whether he thought the U.S. was making a mistake in not speeding up its space program, Eisenhower replied: "Well, no, I don't, because even yet, let's remember this: the value of that satellite going around the earth is still problematical, and you must remember the evolution that our people went through and the evolution that the others went through, from 1945, when the Russians captured all of the German scientists in Peenemünde."

This first myth—that the Sputniks were basically a German, not a Russian, achievement—was quite comfortable for us to believe. Stories to this effect were spread in editorials and cartoons and by word of mouth. Everybody knew that the Germans were good at making mechanical things that worked. The clumsy-handed Russians had simply compelled their prisoners to invent something that could be converted into propaganda.

But the facts are to the contrary. Wernher von Braun and other prominent German rocketeers have told me and others numerous times that the United States got the cream of the Peenemünde rocket talent; the U.S.S.R. got, as one of von Braun's deputies put it, "just the mechanics."

Yet the myth served as the basis for other myths that grew almost in proportion to the Soviet space achievement. One was that the Russians were clumsy engineers. They could build locomotives, bridges, and tractors, but nothing delicate requiring deft precision and true craftsmanship. The beautifully welded and finely instrumented *Sputnik 3* alone would have shattered that myth—had we then been able to marvel at it. When I inspected it in Russia in 1966, it appeared to have been fashioned by a team of master jewelers.

A supplementary myth nourished smug feelings of American superiority by explaining that the real reason the early Russian boosters were so big was that the Soviets had not been smart enough, as we were, to make a small, efficient nuclear warhead that could fit on top of a more modestly designed booster, such as our Atlas. In background briefings to newsmen, Pentagon officers sometimes even referred to the Sputnik boosters as "massive and clumsy." This conveniently rationalized away the Soviet first-generation ICBM booster that lifted the 2,925-pound *Sputnik 3* into orbit on May 15, 1958.

A third related myth was that our smaller boosters forced us to develop superior miniaturized equipment, which gave us a decided long-term advantage. The corollary was that the Russians did not know how to miniaturize equipment. Unknown to us at the time, Soviet scientists were also working overtime miniaturizing their equipment. Soviet scientists, like our own, wanted to conduct as many scientific experiments as possible on each mission. Otherwise, they would have to expend more of their large and expensive boosters. To make full utilization of their scientific payloads, they had to miniaturize equipment, beginning with their first satellites.

The myth that the Russians faked their flights was partly destroyed when Soviet scientists released the first photograph of the back side of the moon. But this myth was soon replaced by another—that the Soviets scheduled space spectaculars purely for propaganda purposes. This is easy to understand. After all, rationalizing the accomplishment of a neighbor by claiming it was done only for show is one of the oldest and most universal of human traits. The Soviets, in fact, were just as guilty of this half-truth as we were. Even Yuri Gagarin referred to some U.S. flights as "spectaculars done for propaganda purposes." There was some propaganda effect, of course, and the record shows that both the U.S.S.R. and the U.S. made the most of their successes. The propaganda motive, however, was not the prime reason for scheduling rocket flights in either country.

After Yuri Gagarin made his historic one-orbit flight in 1961 on what is now known in Russia as Cosmonaut Day, April 12, still another myth presented itself in various forms. Before and after the Gagarin flight, it was said, Russia had lost up to fourteen men in space prior to the announced death of Vladimir Komarov in 1967. One report had it that Sergei Ilyushin, the son of Russia's famed aircraft designer, had orbited the globe three times before Gagarin and had returned to earth so shattered physically and mentally that he had to be permanently hospitalized. Two radio hams from Turin, Italy, the young Judica-Cordilla brothers, claimed to have intercepted the frantic pleas of cosmonauts marooned in space. The Bochum Observatory in Germany was said to have reported the interception of failing cosmonaut pulse and respiratory rates telemetered from a specific point beyond the atmosphere. The Italian newspaper *Corriere della Sera*, which gave the names of those cosmonauts purported to have died in space, listed the names of Yevgeny Andreyev and Pyotr Dolgov. Other reports presented a variety of ways in which cosmonauts had died in space.

When these and other accounts were checked out in the

United States, in Western Europe, and in Russia, there was no conclusive evidence that Russia had lost unannounced men in space, either before or after the Gagarin flight. Ilyushin's reported three-orbit flight preceding Gagarin's one-orbit flight was probably wholly fictitious. The Russian "chief constructor" or "chief spacecraft designer," the brilliant Sergei Pavlovich Korolev, later confirmed that he had personally scheduled an initial flight of one orbit as a safety precaution because of the lack of worldwide Soviet tracking facilities. The North American Air Defense Command's Space Detection and Tracking System in Colorado Springs found there were no objects in space at the location from which the Bochum Observatory and the Judica-Cordilla brothers claimed to have received fading signals. The Russian head of cosmonaut training, Lieutenant General Nikolai Kamanin, said of the Italian brothers: "Either their radios were tuned to a wrong wave length or they suffered acoustic hallucinations." Yevgeny Andreyev and Pyotr Dolgov turned out to have frozen to death on November 1, 1962—not in May of 1961, as reported—during unsuccessful parachute jumps from a Soviet stratospheric balloon. Some Russian cosmonauts, like their U.S. counterparts, did have fatal accidents during training. One was lost in a parachute jump; another died in a plane crash.

Related to the myth of sacrificed cosmonauts was the often-stated generality that Russian space administrators were somewhat careless of human life and were willing to take enormous chances with minimal safety margins. As a matter of fact, Soviet scientists are just as conservative regarding human risks as Americans are. This is shown partly by the relatively long gap between manned flights, the length and intensity of training for a specific flight, and Soviet emphasis on cosmonaut survival training.

Another myth related to cosmonauts is that Russian physical training is so inferior to that of the U.S. that many of their cosmonauts became sick in orbit. A variation of this applied to specific flights—for instance, that the three men who flew in *Voskhod 2* were so sick when they landed that they were driven in an ambulance to a hospital. The facts indicate again that both Russian cosmonauts and U.S. astronauts are splendidly trained physically and mentally, and it is doubtful that any team of doctors and psychologists in the world could possibly rate one group above the other. It is true that cosmonaut Gherman Titov, physician Boris Yegorov, and scientist Konstantin Feoktistov were bothered with periods of spatial disorientation, dizziness, and nausea during their weightless flight, as detailed later in this book. They were not, however, indisposed to the extent that they were incapable of carrying out assigned flight duties or normal post-flight functions.

In 1963, the director of England's Jodrell Bank Observatory, Sir Bernard Lovell, returned from a trip to Russia to start a new line of speculation that was to endure for several years. There are strong indications, Sir Bernard suggested, that the Russians have given up on their program for moon exploration, a program often referred to as the moon race. Despite evidence to the contrary—including repeated Soviet soft landings on the lunar surface, lunar-orbiting probes, and development of a successful retro landing rocket for manned spacecraft—this speculation was repeated and enlarged for several years. One version was that the first step taken by Premier Aleksei Kosygin and Communist Party Secretary Leonid Brezhnev, upon gaining power after Khrushchev's deposition, was to de-emphasize the Soviet moon program by giving the order that the three-man *Voskhod 1*, then in orbit, be immediately returned to earth. Little noted, by those who trade in what people most want to hear, were the enthusiastic words of praise with which Kosygin and Brezhnev greeted the cosmonaut heroes, words that would be difficult later to eat had the new leaders, in fact, drastically departed from policies of that fiery space advocate, Nikita Khrushchev. Also overlooked was cosmonaut Valentina Tereshkova's revelation to a Havana audience that the first Soviet moon team was headed by Yuri Gagarin and that she was on it.

There are also other related myths, such as the belief that the U.S. space program is entirely open while Russia's is largely secret (about half the satellites of both countries are secret) and the persisting belief that the U.S. announces all its failures while the Russians cover up theirs. Actually, the Soviets did conceal most of their failures and aborts at first. But at about the same time that we began a say-nothing policy on our space flights from Vandenberg Air Force Base, the Soviets began to become quite frank about many of their failures, especially lunar and planetary probes.

But perhaps the greatest myth most consistently perpetrated on the people of both of the two great spacefaring nations is that the space goals of "our" country are entirely peaceful and scientific while "our enemy's" purpose is mainly military and therefore sinister. Political leaders and, surprisingly, some scientific leaders of both countries perpetuate this obvious falsehood in almost identical words.

Although the United States has used the more flagrant myths as insulation for its national pride, the primary purpose of this book is not to vanquish myths about the Soviet space program. Nor is the primary purpose to overpraise a program that, because of its nature, has often been disparaged. The primary purpose is to present the nature of a highly significant new scientific thrust beyond the atmosphere that has changed and will change the lives of all of us.

To penetrate and function beyond the panoply of our atmosphere, it is calculated that the Soviets, since the Sputnik breakthrough, have been willing to spend—despite four years of failure of the vital wheat crop and other strong internal pressures to economize—over $20 billion. Gardner Ackley, chairman of President Johnson's Council of Economic Advisers, estimated that the Soviets spent during their first decade about 80 per cent of what the U.S. spent on military and civilian space flights. Soviet scientists and militarists are currently spending on space slightly more than 1 per cent of their gross national product, according to most estimates, at a time when their cost per pound in space is rapidly going down. This is especially remarkable in what Leon Herman of the U.S. Library of Congress calls "the short-blanket economy of the Soviet Union, i.e., the feet get cold when the shoulders get aid."

By the end of the first decade, the Soviets were launching Cosmos satellites so frequently that not even *The New York*

Times always reported them. In addition, they were rocketing highly efficient moon probes, spy-in-the-sky satellites, communication satellites (successful in transmission of color television), meteorological satellites, navigation satellites, multiple-manned satellites, radiation probes, astronomical observatories, and successful deep space probes to the range of the inner planets.

Immediately after Sputnik, an alarmed group of U.S. educators, scientists, journalists, and militarists attempted to point out the magnitude of the Soviet scientific thrust, as it could then be measured. In the first decade after *Sputnik 1* sailed around the world, the U.S.S.R. progressed from 160,000 college graduates in technical subjects in 1957 to over 200,000 (90,000 more than in the U.S.); from fewer than 60,000 science-oriented public libraries to over 400,000 and from slightly over 3,000 research institutions to over 4,600. This proliferation of well-organized scientific institutions was strongly rooted in computer technology; by 1966, over 100 institutions had the word "cybernetics" in their title. By 1967, an estimated 2,000,000 Soviet technicians, soldiers, and scientists were directly involved in their overall space program.

Because of Soviet security, ascertaining the precise nature of this concerted scientific thrust is not easy, but the attempt is important for several reasons. First, we must understand it because we are both competing with and contributing to the exploration of the only noncircumscribed physical frontier ever laid at the threshold of man. No one on either side can really know where this exploration will lead us. There is no doubt that it is significant, or that the Soviet Union contributes powerfully to it. Second, the milestones of that exploration do have an international political effect. "Rightly or wrongly," said General Bernard Schriever, "achievements in space are taken to be a sign of a nation's technological leadership." This thesis was buttressed in the 1960s by an American survey that found the people of 16 out of 21 African nations believed Russia was winning the space race and was therefore the world's leading spacefaring nation. Third, Soviet space operations and capabilities affect our economy.

Whether we like it or not, everything the U.S.S.R. does in space has a direct influence on decisions of Congress, and such decisions radiate through the entire aerospace industry to the taxi driver in Toledo and the cosmetician in Kokomo.

Finally, and perhaps most important, the U.S. must constantly assess the most dangerous part of the Soviet space achievement, the military potential. Like the sea and, after that, the air, space is now becoming the province of fallible man, and as America's first orbiting astronaut, John Glenn, once remarked: "Wherever man has learned to function, regrettably he has also learned to fight." Very little can actually be learned about aerospace technology that could have no military function; the marriage in space of "purely scientific" Venus and militant Mars forms a close union.

There is still another valid reason for examining the personnel, the hardware, and the techniques of the only other major spacefaring nation on our common planet. The last quarter-century has witnessed innumerable scientific breakthroughs, but only two of them can so far be classified as major, profound, and generic. These are the nuclear-energy breakthroughs within the microcosm of the atom and the breakthrough into the macrocosm of space and the infinite universe. Of the two, we know far more about the techniques of atomic fission than we do about the prototype methods of penetrating and examining the cosmos. Indeed, there has been no subject so often on the front pages about which there is less explanatory literature, especially on the Soviet space program.

It is the nature of astronautics, as it is of splitting the atom, to synthesize a great variety of scientific disciplines—from astrophysics to zoolitics. This is why space is also a generic frontier; it affects, for instance, medicine, meteorology, astronomy, genetics, communications, biology, military science, economics, political science, zoology, industrial techniques, jurisprudence, education, geology, and even theology.

The impact of space on man's philosophy might seem, at first glance, to be somewhat remote. Yet a few theologians have already suggested that this ultimately may be the area of greatest effect. Three days after *Sputnik 1* went into orbit,

The New York Times wrote in a perceptive editorial: "The creature who descended from a tree or crawled out of a cave a few thousand years ago is now on the eve of incredible journeys. Yet it is not these journeys that chiefly matter. . . . Rather, it is to understand the heart and soul of man."

The distinguished American church historian Marty E. Marty points out that because of what he calls "military-nationalist fanaticism over hardware," the spiritual side of man was initially unaffected by rampant aerospace technology. Marty regrets that our newest frontier has thus far furnished us "no new Pascal . . . to become the Space Age's first poet or philosopher." But he also wonders about a secondary effect on man's view of himself. "Could it be," he asks, "that the stars were once again working on the humanist imagination, and that once again technology was the agent?"

Theologian Paul Tillich suggested this, perhaps, in *The Future of Religions*, published after his death. Tillich wrote that the reaction to "the breakthrough of the gravitational field of earth was naturally astonishment, admiration, pride. . . . At the same time space flight led to the possibility of looking down at the earth in a kind of estrangement between man and earth." Tillich speculated further that "the possibility of other religiously meaningful histories in other parts of the universe has changed tremendously the cosmic frame of man's religious self-evaluation."

A Russian scientist, Igor M. Zabelin, goes so far as to suggest that the world's population explosion is man's instinctive preparation for resettlement on other planets. "Some inner motivations," says Zabelin, "are leading mankind to new and unknown shores . . . it is gathering its strength."

When the second Russian cosmonaut, Gherman Titov, returned from space, he said, in answer to a newsman's query on what he had found, "I didn't find God out there if that is what you mean." Yet the society that the West regards as godless has found, if not religion, a deep and in some ways spiritual conviction that man has a significant destiny away from his home planet, as suggested by Zabelin. "The earth is the cradle of humanity," wrote the Soviet "father of rocketry,"

Konstantin Tsiolkovsky, "but mankind will not stay in the cradle forever." Tsiolkovsky did much to permeate the Soviet consciousness with what might be called the theory of postdestination. Near his grave at the railroad and machinery town of Kaluga, I found these words of his engraved in stone: "Man will not stay on earth forever, but in the pursuit of life and space will first emerge timidly from the bounds of the atmosphere and then advance until he has conquered the whole of circumsolar space."

Such words as these of Tsiolkovsky are known by virtually every schoolchild in the U.S.S.R. The space monument in Moscow has been erected to his memory. His effect on the Soviet consciousness is far more pronounced now than at any time during his lonely life. And it is an effect that has been consistently misjudged by the West. Western countries, especially the United States, have long believed that "space spectaculars" in Russia were primarily politically inspired and propaganda-oriented—chiefly for external effect. They have believed that Russian attachment to space was a tenuous one and that once the full momentum of the American economy was brought to bear the hard-pressed Russians would have to default on their expensive and precipitous rush across the space frontier.

Nothing could be farther from the truth. The Russian citizen has a highly romantic and idealized concept of his present and future in space. His country entered the space age with a far richer tradition in rocketry than did the United States, and it has responded to the lure of the cosmos—which Tsiolkovsky first phrased in the title of one of his numerous books—somewhat like the proverbial children of Israel responded to the promise held out to them by Moses.

This childlike response, I am certain, is reflected in some subsequent chapters and will undoubtedly strike the reader as somewhat strange, even alien, perhaps. This is due primarily to two things: in the first place, the author has used largely Soviet source material—hundreds of pounds of it—and was influenced to put down fairly directly some of the romanticism and idealization that characterized the Soviet response to space. This response, in other words, is part of the verity and uniqueness of their viewpoint. In the second place, the United States as a nation went through a somewhat similar period of romanticism about a century ago. We feel we have now grown out of that phase and have altered our impression of the world to what we feel is a more realistic and maturely skeptical viewpoint. We are, for better or for worse, more "sophisticated," and this will account for some of the reader skepticism—which first began with the author—to the seemingly childlike prostration of the Russians before the new instant heroes of space.

The degree of Russian emotional-spiritual involvement in space is extremely difficult to judge. But such is the inscrutable nature of the Russian character that it may prove in the long run to be of far more importance than such considerations as economics and external prestige. Those in the West who undertake a realistic assessment of the Soviet potential cannot ignore it. Nor can we ignore Konstantin Tsiolkovsky, who first articulated and lofted the Russians' vision of themselves as a people who had a significant destiny in the cosmos. Lacking the changeable garments of political fashion, the bearded, nearly deaf old man is already fulfilling—along with Sergei Pavlovich Korolev—whatever craving exists in Russia for an enduring national icon.

Some idea of Tsiolkovsky's unique, early position and influence in Russia is suggested by an incident that occurred in 1933, the year the United States, under the Roosevelt Administration, recognized the Russian Government for the first time since the end of World War I. During the time the U.S. was struggling out of its Great Depression and the U.S.S.R. was adjusting to its bloody political purges, thousands of Russians were packed into Moscow's cavernous Red Square for their traditional May Day observances. Tsiolkovsky, who was then 76, was too weak to be there, but near the end of the ceremonies the masses grew silent as his radio voice was piped in from his home in Kaluga. Amplifiers boomed his weak voice out over the sea of faces.

"Now comrades," he said, "I am fully convinced a dream of mine—space flight—for which I have sought and found a theoretical foundation, will come true. For 40 years I have been working on the rocket motor, but I thought that a journey to Mars could take place hundreds of years later. Time, however, moves quicker, and now I am sure many of you will be witnesses to the first transatmospheric flight. Heroes, men of courage, will inaugurate new airlines: Earth-Moon orbit! Earth-Mars orbit! Moscow-Moon! Kaluga-Mars!"

This impassioned and romantic exhortation would have sounded strange, indeed, to American ears at the time the song "Happy Days Are Here Again" could be heard on every radio. And it may have sounded strange, also, to some Russian ears at a time when motion pictures with sound were just being introduced in the Soviet Union.

But to the majority of Russians, it was the first articulation of a dream to which they were peculiarly susceptible, a dream that—within their lifetime—was to become very real. Tsiolkovsky not only bequeathed to his country the theoretical basis and formulas for rocket flight, but also kindled the national imagination and thus erected a firm launch platform for the powerful and unexpected surge across the frontier of the cosmos.

2
The Trailblazers

WHEN I visited Kaluga in 1966, Russians, including women workers in soiled denim and muddy scarves, were already erecting Tsiolkovsky's cathedral, actually an ultramodern space museum overlooking the verdant Oka River valley and bordering forest. In an adjoining park, where Russian poplar trees sifted down their "summer snow," were the monument and inscriptions that Yuri Gagarin and other cosmonauts saw in their postflight pilgrimages to Kaluga.

Konstantin Edwardovich Tsiolkovsky, who died in 1935 at the age of 78, was engineer and prophet, inventor and idealist. Many of his numerous aphorisms, such as "What is impossible today will be possible tomorrow," are well known throughout the Soviet Union. Most important, unlike such other national figures as Stalin and Khrushchev, he was completely apolitical. Whatever fate befalls the evanescent, political god-heroes of the Soviet Union, no Russian can ever successfully dethrone the man from Kaluga who pointed out goals never dreamed of by Marx and Lenin.

"In my imagination," Tsiolkovsky wrote of his childhood, "I could jump higher than anybody else, could climb poles like a cat and walk ropes. I dreamed that there was no such thing as gravity."

At the age of nine, a severe case of scarlet fever rendered him almost totally deaf. He could no longer attend school. "My deafness," he wrote, "made my life uninteresting, for it deprived me of companionship, of the possibility to hear people around me, and to imitate them. My biography is poor in human contacts and conflicts." He later admitted that the years between the ages of 10 to 14 were the darkest and saddest of his life. He was apparently lost in the confused silent world of sensory deprivation at a time his mind should have been most responsive. "I try to recall those years but I cannot think of a single event," he wrote.

But his father's library and a penchant for invention brought the youth back to life. Between fourteen and sixteen, Konstantin devoured books on mathematics and natural history, energetically constructed all kinds of objects, ranging from tissue-paper balloons to a turner's lathe, experimented with a primitive range finder and an astrolabe (an instrument that Christopher Columbus used in his ocean crossing), and built his own height finder, which he used to determine accurately the height of a distant fire tower without going outdoors. "This made me believe in theory," he explained.

His parents decided their precocious son should continue his self-education in more appropriate surroundings and sent him, at sixteen, to Moscow. In the capital city, on an allowance of only fifteen rubles a month, he lived almost entirely on brown bread in order to purchase small quantities of quicksilver, sulfuric acid, books, and equipment for his experiments. Every three days, he went to the baker's and bought nine kopeks' worth of bread. "My diet," he recalled, "did not damp my spirits."

It is not known precisely how, in this odd chrysalis of noxious acids, homemade apparatus, and bread crusts, the idea of space conquest first occurred to him. But at one point he seized upon the notion that man could leave the earth with a simple motor utilizing centrifugal force. He invented a sort of box into which were suspended two oscillating pendulums. Balls attached to their upper ends described circular arcs in their movements, and Tsiolkovsky thought their centrifugal force would cause the box to rise all the way to outer space. His box so gripped his fancy that he physically shook as he wandered the streets of Moscow all night long thinking, as he put it, "of the grave consequences of my invention."

The obsessed youth must have cut a startling figure with his uncut hair and his yellow-stained and acid-eaten trousers, which looked, according to one observer, as if mice had got to them. But by dawn, his scientific logic had prevailed over his soaring fantasies, and he painfully admitted to himself the futility of his invention. But he never fully detached himself from the emotional exhilaration born of an ultrasimple solution to a highly complex problem of physics. "My heart still swells," he wrote later, "with the exultation I experienced on that unforgettable night . . . From that moment, the idea of space flight never left my mind and urged me to study higher mathematics."

The questions he now began to ask himself indicate the restless range of an independent, nineteenth-century mind. Can the energy of the earth's motion be utilized for practical purposes? Can a train be made to run along the equator freed from gravity by centrifugal force? Can an airtight aerostat be constructed capable of floating in the air for an indefinite time?

After three years in Moscow, he passed an exam to become a "people's school teacher" in a village near Kaluga, but his main interest was in his weird new homemade lab. His left-handed humor was expressed through several machines that could grab unwary visitors in an octopus-like grip, cause their hair to stand on end and sparks to fly from their bodies. The scientific prank was harmless, as perhaps only the delighted amateur scientist knew.

His first formal scientific recognition came as a result of a second paper he sent at the age of twenty-four to the Society of Physics at St. Petersburg. The paper won him unanimous election to the society. Two years later, he wrote a diary-type book called *Free Space*. The diary, written in the 1880s, imaginatively examined the characteristics of objects and motion in what we now know, firsthand, as the familiar state of weightlessness. One sentence in the entry of February, 1883 is particularly noteworthy: "Consider a cask filled with a highly compressed gas. If we open one of its taps, the gas will escape in continuous flow . . . the result [in free space] will be a continuous change in the motion of the cask." He continued to work alone, with little contact with the outside world of science. Often he did not know whether something had been done before, and he had to prove out each project or experiment from the ground up. His time may occasionally have been wasted, but he gained valuable perspective on the whole of a given scientific problem. His judges were the high scientific priests at St. Petersburg, to whom he regularly sent the results of his experiments. "At first," he wrote, "I discovered old truths, then some not so old and finally some quite new."

By the time he was twenty-eight, he was concentrating almost entirely on a theoretical proposal for an all-metal aerostat (the term "dirigible" came into use much later). His invention called for the hydrogen gas to be ingeniously heated by the engine exhaust. Its volume could be changed without altering its exterior shape. He drove himself for two years on this project and wrote a book on it during the early hours before he left each morning for his teaching duties. But the only results of his labor were scattered favorable comments from some of the scientists at St. Petersburg. His request, in 1891, for a grant to construct a working model was denied. It was not until 1895 that Germany's Count Ferdinand von Zeppelin completed his dirigible design, and not until 1900 that he completed his first experimental lighter-than-air vehicle.

Tsiolkovsky continued to work. He constructed a large mechanical vulture that may have flown well enough to impress the children of the town. Once, according to his grandson, he built a wood fire under a paper balloon attached to a string. When the string caught fire, the balloon escaped, dropped live cinders on the town, and finally landed on the roof of a shoemaker. The shoemaker, so the story goes, was so mad he would not give back the balloon.

For virtually the remainder of the nineteenth century, Tsiolkovsky continued to work on dirigible plans and an advanced design concept for an all-metal, streamlined single-winged airplane. His experiment in building the mechanical vulture with flapping wings had convinced him that "technically, it is very difficult to imitate the bird, owing to the complex nature of the movements of the wing and tail." His proposed design, instead, was shaped like a "soaring bird in place of whose head we must imagine two propellers turning in opposite directions."

Soviet historians overemphasize the superiority of Tsiolkovsky's design to that of the Wright brothers of Dayton, Ohio, who made their first airplane flight at Kitty Hawk, North Carolina, in 1903. It is futile to compare the two, because Tsiolkovsky's design—remarkably advanced though it was—existed only in the form of drawings, intricate calculations, and a single article, "The Problem of Flying by Means of Wings." Tsiolkovsky well understood the laws of aerodynamics, however, and constructed the first wind tunnel in Russia.

Most of Tsiolkovsky's rocket research was done in the workshop of a small wooden house located on an inclined street leading down to the Oka River. When I visited it and talked at length to his grandson, Alexei Kostin, it was possible to see a number of Tsiolkovsky's models and tools, his pioneering wind tunnel, and a dozen or so of the small early Soviet experimental rockets based upon his theories.

His workshop was a glassed-in rectangular room, located on the second floor, where he also preferred to sleep away from other members of the family. Beside his bed are two homemade ear trumpets, shaped like tin funnels, which he used to say "was one of my inventions that worked." When German troops occupied Kaluga in World War II, they used some of his furniture for firewood, but his prized workbench —which he skidded into Kaluga one winter day on a sled—is still intact. A small door leads from his workshop onto the sloping roof of a shed. He was so fond of spending evenings on the shed roof looking at the stars that his family affectionately dubbed his exit as "the door to outer space."

Tsiolkovsky moved to this house in 1892, about the same time he started serious work on rocket propulsion. "For a long time," he wrote, "I thought of the rocket as everybody else did—as just a means of diversion and of petty everyday uses, but, gradually, stimulated by the fantasy writer Jules Verne, I began to investigate its possible use in space flight." His first article on rocketry, published in *Scientific Review* in

1903, substantiated the use of liquid-propelled rockets for space flight—based on the elementary laws, which he knew so thoroughly, of physical mechanics. "Let us imagine a missile as follows," he wrote, "a metal elongated chamber supplied with light, oxygen, absorbers of carbon dioxide, miasmata, and other excretions, and intended not only for containing different physical instruments but also for carrying a man controlling the chamber." He envisioned that the rocket would use a mixture of a propellant and liquid oxygen as an oxidizer. His design called for the propellant mixture itself to cool the walls of the combustion chamber (a procedure now widely practiced and known as "curtain cooling"). Attitude control would be by jet vanes and a control nozzle.

In his later years, he improved the basic design and experimented with kerosene fuel. Kerosene, known as RP-1, is now a widely used rocket propellant. He also made the necessary calculations to work up the practical theory of the "multistage rocket," or what he sometimes called a "rocket train."

In developing his theory of rocket flight, his mind foresaw and took into consideration the forces of terrestrial gravity and drag. His formula for overcoming these forces formed the basis for rocket dynamics and came to be known as Tsiolkovsky's formula.

At Kaluga, at the same time when he was concerned with the practical, down-to-earth calculation, he allowed his restless mind to soar off in futuristic fantasies. His preoccupation with laboratory experiments was matched by his remarkably imaginative exploits of science fiction. He wrote over a dozen works of science fiction, including the most presentient ones, *On the Moon* and *Beyond the Earth's Atmosphere*. In *On the Moon*, he realistically visualizes himself and another man waking up in their own house, which has been relocated on the lunar surface. "When we wanted to move rapidly," he wrote, "we had first to lean perceptibly forward like a horse when starting to pull a heavy cart. . . . Every movement we made was exceedingly light. How boring to descend a staircase, step by step! How slow to move at a walking pace! Soon we abandoned all these procedures, so suitable on the earth but so ridiculous here. We learned to move by leaps and bounds; we took ten or more steps at a bound up and down the staircase like the most harum-scarum schoolboys. Sometimes we jumped the whole flight or leapt through the window. In short, by force of circumstances we were transformed into jumping animals, like grasshoppers and frogs.

"So, after rushing all over the house, we jumped out and ran toward the nearest mountain. The sun was cadaverous and dazzlingly bright and seemed almost blue. We could see the stars and planets, which were also, for the most part, tinted blue. Neither stars nor planets twinkled, which made them like silver-capped nails studding the black firmament."

Seldom did his fantasies depart from the truths of science, as then understood. In fact, his science fiction seemed to complement his more serious obsessions. "First," he wrote in 1926, "inevitably the idea, the fantasy, the fairy-tale. Then scientific calculation. Ultimately, fulfillment crowns the idea."

"Many times," he later wrote, "I essayed the task of writing about space travel but wound up becoming involved in exact computations and switching to serious work."

Even at the age of 78, he confessed to "devoting half my time and energy to the problem of overcoming terrestrial gravity and making flights into space." Almost two entire decades of his life were taken up with the theoretical development of the rocket engine and the jet engine. He envisioned the use of space suits, the mooring of people and objects in the weightless state, the closed ecological system (food-producing gardens), rendezvous and docking, and many other aspects of space exploration that are now becoming routine.

Before he died in 1935, he received a rare tribute from the distinguished German rocket pioneer Professor Herman Oberth, whose own monograph, *The Rocket into Interplanetary Space*, was to become the bible of the rocketmen at the German Baltic town of Peenemünde.

"You kindled this fire," Professor Oberth wrote to him. "We shall not let it die; we shall try to realize man's greatest dream."

One day in 1959, I talked briefly with Professor Oberth at

what is now called Cape Kennedy. As we talked, Oberth's tired old pale-blue eyes returned time and again to the great rockets and rocket gantries perched on the palmetto flatlands of the Cape. Much of the dreams of Konstantin Tsiolkovsky, of the American, Robert Goddard, and of Herman Oberth then stood around us, poised for the future. Neither Tsiolkovsky nor Goddard ever visited the bristling ranks of the Cape's mighty rockets, but it was a view of man's destiny they actually had seen many times in the obsessed eye of the mind.

By the time of his death in 1935, Konstantin Tsiolkovsky was a popular as well as an academic hero of the Soviet Union. Most of his writings had been published at public expense, and he was financially independent, thanks to a state research subsidy. On his seventy-fifth birthday, Stalin himself sent a telegram of congratulations to the special birthday ceremony that had been arranged by the *Osoaviakhim* (Society for the Promotion of Defense and Aerochemical Development). He was an object of highly complimentary publicity in the Soviet press, and, of more importance to him, was frequently consulted by the U.S.S.R.'s growing roster of amateur and theoretical rocketeers.

His reputation had not yet spread to America, but the German Oberth had known of his work for a decade. Oberth, who had written his now-famous booklet in 1923, was pained to learn that Tsiolkovsky in his old age considered him a rival and a "latecomer." The German once wrote the ailing Russian. "I would be the last to dispute your priority and your merits in the field of rockets. My only regret is that I had not heard of you before 1925. Had I known your excellent work, I would surely have been further along in my research now, and I would have spared myself much unnecessary labor . . . you lighted the light, and we will work until mankind's dream comes true."

But this was small mollification to the aging Russian, who steadfastly refused to share credit with either Germany's Oberth or America's Robert Goddard. Tsiolkovsky's decades of silent torment and his long struggle for recognition had left the sullying mark of bitterness upon him.

But if Tsiolkovsky failed to receive what he considered the rightful homage of all mankind, his reputation in Russia was such that it eclipsed, for a time, other important Russian space scientists and engineers who were his contemporaries. Among the earliest of these was the brilliant theoretician Ivan Meshchersky, who along with a devoted Tsiolkovsky disciple, Friedrich Tsander, and the even earlier theoretician, Kibalchic, helped give the Soviet Union an enviable early background in astronautics. All of these men now receive prominent but subordinate credit in the Russian space museums and libraries at Moscow and Kaluga. Meshchersky performed important early work on the dynamics of bodies of variable mass, and the old revolutionary Kibalchic, who died in 1881, worked out the theory of using rocket motors to guide an interstellar spaceship while he was a political prisoner of the Czarists. Because of his youth and enthusiasm for building things, it fell to Friedrich Tsander to convert the early theories, especially Tsiolkovsky's rich legacy, into the first functioning hardware. Tsander, who named both his children after stars, was among a handful of rocket enthusiasts who read every word Tsiolkovsky wrote and who spearheaded the Soviet practical work in rocketry—especially on rocket motors—that began in the late twenties. As early as 1924, Tsander made a detailed proposal for a spaceship, and in 1929 he presented to Moscow scientists a large model of his proposal.

The present era of Soviet enthusiasm for rocket-hardware development began, according to Professor G. Petrovich of the Soviet Academy of Sciences, on May 15, 1929, not in Moscow but in Leningrad.

"On this day," says Petrovich, "we started the experimental development of electric rocket motors and liquid-propellant rocket motors at Leningrad's Gas Dynamics Laboratory [GDL]." Essentially this same group of now-elderly men are presently part of the senior staff of the heart of the present Soviet space effort, the Research Design Office (known as the OKB) in Moscow.

"Thus our experience," says Petrovich in reviewing the several decades, "ranges from the first experimental motors with a thrust of several kilograms to rocket motor installations, the power of which is measured in tens of millions of horsepower."

This early group of enthusiasts, the members of the tiny GDL sublaboratory of a large military base, worked for the next four years on both electric and liquid rocket motors. For fuel on the experimental electric motors they used either metallic wires or electroconductive liquid jets of mercury or various electrolytes. The fuel was fed continuously into a chamber equipped with a nozzle, where it was exploded by an electric current. In a way, they were putting the cart before the horse, as the electric motor does not work efficiently until it has reached high altitudes. Experiments with the electric motor ceased in 1933, when the requirement for first raising it to altitude shifted emphasis to the development of a promising, Tsiolkovsky-based liquid rocket that had first received design attention in 1930. The first liquid engine successfully fired in the laboratory was 2 feet long and produced a thrust of 20 kilograms. It can now be seen in the space pavilion of the Economics Achievement Exhibit in Moscow. The first working liquid model engine, the ORM-1, was completed in early 1931 and used the very rare nitrogen tetroxide and toluene as fuel components. Since the budget was limited and the expensive nitrogen tetroxide was virtually unobtainable, the GDL group diverted their efforts to trying to produce their own fuel. They also, as early as 1930, experimented with motors that used nitrogen oxide as an oxidizer, because it was readily available and could remain in a liquid state through a wide temperature range. Also, as early as 1930, the GDL group used cryogenic oxidizers (liquid oxygen) during static tests, as well as tetranitromethane, perchloric acid, and hydrogen peroxide. By 1931, they had successfully tested at least one hypergolic (self-igniting) fuel as well as a pump fuel powered by the combustion chamber.

Coming at these early dates, the variety of pertinent laboratory tests is truly astonishing, although the Soviet Union had not yet successfully launched a rocket. A comparision with the situation at that time in the United States and in Germany is revealing.

In the U.S. by 1930, Professor Robert Goddard's classic treatise, *A Method of Reaching Extreme Altitude*, was eleven years old, and he had already launched several liquid-propelled rockets. Among these was an 85-pound rocket he named *Nell*, which rose 7,500 feet above the New Mexico desert at nearly the speed of sound. Goddard and his crew of about five helpers, including his wife, Esther, worked essentially in isolation. The U.S. military had scorned his efforts, and the only outside source of encouragement was the Smithsonian Institution and Colonel Charles Lindbergh who, just two years after his famous transatlantic flight, developed a personal and long-sustained interest in the work of the balding professor at Clark University, Worcester, Mass. Lindbergh talked the American millionaire and philanthropist Daniel Guggenheim into providing $25,000 annually so that Goddard could carry on his rocket experiments.

Unlike his Russian or German counterparts, Goddard worked entirely alone. There was no opportunity to develop a nucleus of knowledgeable enthusiasts who would form—as on the Continent—the valuable cadres of the future age of space. Until his death in 1945, the only use the military ever made of the talents of Robert Goddard was in the development of JATO rockets to assist military aircraft on takeoff.*

The situation in Germany was more like that in the Soviet Union. By 1930, Oberth's "space bible" was seven years old and had influenced the formation in 1927 of the famous VER, Society of Space Travel, in Breslau. Rocket-motor development had progressed to the point where a solid-propellant rocket car, driven by Kurt Volkhart, had already performed at Ruesselsheim, Germany, and amateur rocketeers and children all over Germany were firing "toy" rockets. In 1930, the VER moved to an airfield outside Berlin for rocket testing.

---

* The military's early neglect of the significant Goddard experiments was partially compensated for when, years after his death, the U.S. awarded his family $1,000,000 for use of his numerous patents.

Among its members was a young, firm-jawed, handsome technician named Wernher von Braun and a young doctor, Captain Walter Dornberger, who persuaded the German Army to donate 5,000 marks (about $1,200) to bolster the sagging test activity. The German nucleus was poorly financed, but it was organized and, most important, it was gaining the invaluable experience of a highly varied series of actual rocket tests.

The pre–World War II Russian rocket societies can swap story-for-story with the Germans on their early efforts to obtain enough funds and materials for research and test programs. In the early thirties, the GDL activity at Leningrad was expanded with the creation of two new organizations under *Osoaviakhim*: GIRD in Leningrad and MOSGIRD in Moscow. The Leningrad organization was led by N. A. Rynin, author of an important nine-volume treatise on interplanetary communications, and Dr. Yu. I. Perelman whose chief function was to popularize astronautics. The MOSGIRD organization included as its deputy the brilliant Tsiolkovsky student Friedrich Tsander and three other men that later became highly honored in Russia: M. K. Tikhonrarov, Leonid Korneyev, and Y. A. Pobedonostsev. Tikhonrarov, in fact, may have subsequently become the never-named, mysterious "Chief Theoretician" of the Soviet Union, and a gifted and still unpublicized engineer, I. P. Fortikov, is believed to have been the first head of MOSGIRD. Another member who is the most likely candidate for the never-identified "Chief Engine Designer" was V. P. Glusko.

In the formative years, the government-sponsored group worked in a primitively equipped basement of an apartment building at 19 Sadovo-Spasskiy Street, near the center of Moscow. Initially, the members possessed only two half-broken lathes and none of the measuring devices so essential to rocket work. When they needed silver for soldering, they often had to bring a silver cross, a teaspoon, or a drinking cup from home. Like Goddard, Oberth, and Tsiolkovsky before them, they went through a baptism of neglect and ridicule, which was only slightly moderated. Food was rationed in Moscow in the thirties, and members were once refused ration books by local bureaucrats on the grounds that they were nonworking people occupied with "nonsensical fantasies." Members, who jokingly called themselves "the engineers who work for nothing," had frequently collected money among themselves, as much as each could afford, in order to buy even the most rudimentary equipment.

It was under these conditions, where minimum facilities were balanced by only application and enthusiasm, that the Soviet Union got its first rocket. But not before one experiment blew up accidentally when a rocket propellant ignited in their underground "cosmic laboratory," as they called their workshop. One member recalls that the apartment building above them was engulfed in smoke as he tried to explain to the Moscow fire brigade what they were doing.

Nevertheless, on November 25, 1933, in a wooded area near Moscow, the Russian experimenters launched a Tsander-designed brass and aluminum rocket 18 inches long. The tiny rocket engine, named 09 (sometimes called OR-2), used compressed kerosene and liquid oxygen to produce a power of 52 kilograms. The engine burned for 18 seconds. Tsander himself did not witness those few important seconds of history; he was dying in a nearby sanatorium at the youthful age of 46.

Within a few years, liquid rockets designed by Glusko were also launched. In 1933, Glusko's ORM-52 rocket, burning kerosene and nitric acid, produced 110 pounds of thrust, enough to attract the attention of aviators looking for an additional means of speeding and lifting heavily loaded airplanes on takeoff. By the mid-thirties, a Tikhonrarov-built rocket reached an altitude of 6 miles (Goddard's highest rocket had then gone 1⅓ miles) and Rynin's nine-volume *Encyclopedia of Astronautics*, the first in any nation, had wide distribution in Russia.

For the next few years, Russian and German progress was apparently neck and neck. A von Braun-developed rocket of 650 pounds thrust, for instance, was launched just four months before Glusko launched his ORM-52 rocket of approximately the same power in 1933.

One symptom of the military application of science in the

Soviet Union is the sudden disappearance from open scientific literature of a previously concerted line of research. From the mid-thirties on, as the Soviets began to find potential military uses for the small rockets then in existence—either as artillery or aircraft takeoff augmentation—information on Soviet astronautics grew thinner. Evidence that the Kremlin was taking note of the work of GIRD and MOSGIRD was the establishment in 1933 of the Reactive Scientific Research Institute in Moscow. For the next half-dozen years, this new group, which drew heavily on GIRD and MOSGIRD leadership, concentrated on liquid-propellant rockets and small jet-assisted takeoff rockets. The government allowed a brief announcement of the successful firing of the first two-stage rocket near Moscow in 1939. Then, as World War II erupted, Soviet efforts were blanketed under military secrecy. The Russians regarded Nazi Germany as their chief enemy then, and were well aware of the German Army's active financing of rocket development under the civilian Wernher von Braun and Army officers Captain Walter Dornberger, Major von Horstig, and Colonel Karl Berker.

After about 400 test firings of the Russian rocket RD-1, it was used against the Germans on such combat aircraft as the PE-2, the LA-7R, and the YaK-3. The RD-1 thrust of 300 kilograms was superseded by the RD-2, with 600 kilograms thrust. By 1945, the RD-3, which was equipped with a gas generator and a turbine pump installation, was static tested at 900 kilograms thrust, according to Petrovich.

World War II rocket research and employment of booster rockets, air-to-ground armaments and the development of jet-engine propulsion* in the general field of military aviation was highly significant technically, but the widespread Russian adoption of solid-fueled rockets as fast-firing field-artillery pieces was the real popularizer of the rocket in Russia, especially in the ranks of the Red Army. On July 28, 1941, one month after Hitler invaded Russia, the Supreme Command of the Red Army adopted all missiles for military ap-

---

* The Jet Propulsion Research Laboratory was established in Moscow in 1933.

plication. The artillery rocket was the almost immediate result.

When the German divisions swept deep into Russia, they were often preceded by a devastating artillery barrage, which in some cases consisted entirely of rockets. The Russian field troops who faced the sustained German *Wurfgerat* and *Nebelwerfer* rocket barrages had a healthy respect for them and asked for similar support from their own officers. To their relief, the Red Artillery already had such units, in addition to what they called their "Mother Cannon," on which they had traditionally relied to repel invaders. Fortunately for the Red Army, it was soon in a position to requisition thousands of small, solid-propellant artillery rockets and to order others from the U.S. The first mass use of the rockets helped decimate the surrounded armies of German General von Paulus. Since the rockets were launched from a simple rack, virtually any type of vehicle could be used as a launcher. The American writer Martin Caidin reported in 1963 that at least some of the mobile launchers used against von Paulus were U.S. Lend-Lease Studebaker trucks firing, in some instances, Lend-Lease rockets shipped from Tennessee factories.

The most famous of these Russian artillery rockets were the rapid-firing "Katyusha" rockets, used to help turn the tide of the war at Stalingrad. The Katyushas had an effective range of 10 kilometers, and the massed launch racks of "Stalin organs," as they were often called, could bombard the Germans with literally thousands of detonations per minute.

Although the Russians had an operational turbojet by the end of World War II, they had not yet developed a major "cruise" rocket like the German V-1 or the far more advanced ballistic V-2 (known to the von Braun group that developed it as the A-4) that rained down on London and Antwerp.

This grim, curtain-raising chapter of military rocket development belongs strictly to the Germans. Since a great deal of controversy has surrounded the question of how much help the Germans later gave either the Russian or the U.S. rocket

THE TRAILBLAZERS 31

programs, a brief look at the death of the great German rocket base at Peenemünde should set the record straight. I spent five days interviewing former Peenemünde Germans at Huntsville, Alabama. All of them agreed that the U.S. received the lion's share of German talent and experience in rocketry. As one of them put it, "The U.S. got the brains; the Russians got the mechanics." One Peenemünde veteran, Dieter Huzel, now works on the U.S. Saturn rocket program at North American Aviation in California. In his book, *Peenemünde to Canaveral,* he gives a fascinating blow-by-blow account of the rise and fall of Peenemünde.

As early as August, 1936, German Air Force General Kesselring and Army Colonel Berker ordered ground broken for a rocket-development center on the windswept Baltic coast. This was to become the famous Peenemünde that, under the wartime pressure of Hitler's frantic search for a "wonder weapon," broke through the atmosphere with long, slender, powerful and deadly "cucumbers." The first A-4, in fact, according to one of the Peenemünde veterans at Huntsville, successfully passed its static test on April 18, 1938, seventeen months before Hitler's panzer divisions opened World War II by sweeping into Poland.

By the spring of 1944, the first German-fired V-2 landed on foreign soil—in Sweden. But it was so damaged by the impact that the Allies could not then guess what was to come. By September 6, the first of many tactical V-2s was launched against England. By war's end, over 3,000 V-2s had bombarded London. If they had become operational just one year earlier, they could well have reversed the outcome of the war in Europe. The V-2 was a true ballistic rocket and the largest ever constructed. Technically, it could have been made even larger, but it had been deliberately restricted to its 46.2-foot length and 5.4-foot diameter so it could be transported by truck on Germany's superhighways and field roads. Propellants were 75 per cent alcohol, plus liquid oxygen. At sea level, it developed 56,000 pounds of thrust for 65 seconds, sufficient to lift its 28,500 pounds of takeoff weight. After it reached a velocity of 3,600 miles per hour, its engine cut off and its 2,200-pound warhead coasted through a maximum altitude of about 60 miles to strike its target 200 or more miles from its launch pad. Although the early V-2s were far from fully reliable, they were eventually produced in great quantities. Wernher von Braun estimates one plant alone produced 300 V-2s a month.

It would be difficult to estimate the effect when the full impact of this remarkable new weapon was felt in Allied high-command circles. Since once launched it could not be stopped, its immediate strategic superiority was overwhelmingly apparent. It was the subject of the highest levels of discussion and planning. Fleets of bombers were immediately redirected to Peenemünde and suspected V-2 launch sites. From the halls of the Kremlin to the corridors of the Pentagon, there was grudging recognition that the common enemy had produced that most remarkable military rarity, an effective and reasonably accurate weapon against which there was no known defense.

Since the English were within range, they felt its startling impact most immediately, and the bombers of the Royal Air Force flew urgently and repeatedly against the technological citadel of Peenemünde—sometimes exploding their bombs in wet concrete, since construction continued almost until the German surrender. One massive raid alone involved 600 RAF bombers. The raid of August 17, 1943, killed 815 people, half of whom were Russian war prisoners, according to a postwar U.S. Army intelligence report. Although most of Russia and all of the U.S. were out of range of the V-2, both countries suspected that larger rockets were under development. They were right. At the time Wernher von Braun and his team fled westward to escape the advancing Russians, Peenemünde not only had under development a large rocket capable of being launched from an aircraft at 26,000 feet, but also had detailed plans for a practically conceived A-10 and A-9 rocket combination with a calculated range of 2,600 miles. This, the world's first detailed blueprint for an intercontinental ballistic missile, could have reached Moscow.

Some of this awesome new potential of modern warfare must have motivated the high commands of both the Russians and the Western Allies as the two great advancing armies slowed just short of a highly delicate confrontation. Somewhere in between, in the no-man's-land of a defeated Germany, lay the brains of Peenemünde, considered by some to be the richest war booty of modern times.

3
The Treasure Trove of War

When the Russians reached the great Baltic rocket base of Peenemünde they found it devasted far beyond the considerable damage inflicted by the repeated raids of the RAF. The "brains" of Peenemünde had seen to that. Before von Braun left, he called his civilian staff of rocket experts together to plan the destruction of Peenemünde and their flight to the West. Amid the general gloom as the Russians inched steadily closer, Dieter Huzel reports, von Braun gave them all temporary relief when he firmly announced: "This is important. We will carry our administration and structure straight across Germany. This will not be a rout." According to Huzel, there was no doubt in anyone's mind that when they surrendered, they would surrender to the Americans, not to the Russians. Von Braun first ordered Dr. Kurt Debus, later to be put in charge of all NASA launches at Cape Kennedy, to cease all firings at Peenemünde. Then the remaining Germans—almost all of them civilians—stripped their files, burned some records and packing the most important in carefully coded and indexed cartons and boxes that eventually filled five trucks.

"These documents," according to Huzel, "were of inestimable value. Whoever inherited them would be able to start in rocketry at that point at which we had left off, with the benefits not only of our accomplishments, but of our mistakes as well—the real ingredient of experience. They represented years of intensive effort in a brand-new technology, one

THE TREASURE TROVE OF WAR

that, all of us were still convinced, would play a profound role in the future course of human events."

As von Braun and the cream of the Peenemünde talent fled southwestward toward the mountain resort of Oberammergau, Huzel followed with his precious five trucks containing what he called "a cache of scientific documents unlike any in history." After innumerable delays, Huzel finally unloaded his cargo into a remote Harz Mountain cave near Dörnten, Germany. Then he dynamited the cave entrance. When Huzel rejoined the Wernher von Braun group, von Braun sent his brother, Magnus, to Reutte to surrender the group en masse to the Americans there. This initial group consisted of about 150 of "the best scientists and technicians," according to a U.S. Army intelligence report released years later. It included, among others, General Walter Dornberger, commander of all German rocket forces; Hans Lindenberg, V-2 combustion chamber engineer; Bernhard Tessman, chief test designer at Peenemünde; and, of course, Peenemünde's engineering director, Wernher von Braun himself. Such experienced men as Dr. Kurt Debus, Peenemünde's launch director, and Albert Zeiler later joined them. Eventually, over 300 rocketmen and six shiploads of rocket equipment, including the entire contents of the secret Harz Mountain cave, which was reopened in April of 1945, were shipped to the United States. The cave trove went directly to Maryland's Aberdeen Proving Ground for analysis.

In the meantime, the Second White Russian Army under General Konstantin Rokossovsky had overrun what was left at Peenemünde on May 6, 1944. A handful of German technicians were captured. The Russians were no doubt disappointed at the meagerness of the intellectual and material booty they found there and, later, elsewhere in Germany. But they were far from empty-handed. There were three V-2 plants in Germany. In July of 1944, the Russians captured the Eastern Plant near Riga, which gave them some salvage opportunity. They also captured the rocket range at Leba in Poland, the rocket fuel pump manufacturing center at Gorlitz, and the rocket production lines at Breslau.

Most former Germans now working and living in Huntsville admit that the Russians did get one important Peenemünde electronic expert, Helmut Groettrup, who chose to surrender to them. But they insist that, with this exception, the Russians got none of the brains at Peenemünde. Elsewhere in Germany, the Soviets got such talent as ballistics expert Dr. Waldermar Wolff and, eventually, several other knowledgeable scientists, including guidance specialist Dr. Wilhelm Fischer.

A number of the present Huntsville Germans charter a plane to Europe each summer and on these visits occasionally talk to former Peenemünde technicians who were taken by the Russians and later released. These sources agree that Germans were used by Russia only to assemble and launch the V-2s that were immediately test-fired in Russia. "They were kept in the dark about other aspects of Russian rocketry," says Albert Zeiler. "All they saw were our own cucumbers. As soon as the Russians picked up the V-2 know-how, they sent them packing."

In 1958, long after he had talked with German rocketmen who had briefly served the Russians, von Braun testified in Congress that "German scientists taken into Russia were obviously not even aware of the large and extensive ballistic missile program that was going on inside Russia." He further testified: "They did not, to any appreciable extent, actively participate in the hardware phase of the rocket and missile development program in Soviet Russia."

The German surrender, in effect, simultaneously gave three nations an adequate number of the astonishingly advanced V-2s on which to experiment and build. Even if a nation had virtually no tradition in rocketry, as was the case in England, the V-2 booty—with its painstakingly evolved high-pressure propellant pumps, heat-resistant combustion chambers, and delicate gyroscopic guidance—would have served as a more than adequate base on which to build the new science of astronautics. The V-2, in essence, was the embodiment of all the major conclusions of Tsiolkovsky, Oberth, and Goddard.

In Britain's case, as soon as officers became aware of the true nature and might of the "miracle weapon" that rained down upon them, they formed a special group under the designation "Operation Backtrack" to salvage V-2 parts and rare in-flight photographs and to try to trace the new weapon back through its launch facility to its mechanical genesis. By war's end, this efficient and desperation-born organization had almost reconstructed both the principle and the mechanical configuration of the V-2. When the British received captured V-2s, they were quickly able, with some help from their new German prisoners, to prepare and fire a number of V-2s from England.

England, Russia, and the United States thus began the second great phase of rocketry as comparative equals—with important exceptions. Russia already possessed a well-organized cadre of theoretical and practical rocketmen, while neither England nor the U.S. had such a scientifically curious, well-motivated, and well-rounded group. England was already feeling the severe economic effects of a tremendously expensive war and was preoccupied with restoring her tight, partially decimated little island and salvaging a disintegrating empire. The United States had demonstrated to the world an amazing technical virtuosity and superiority by dramatically ending the Pacific War with a formidable wonder weapon of its own—the atom bomb. At war's end, Russia, of course, had no secret weapon, no atom bomb, no rocket—not even the V-2—with a range sufficient to reach what it came to regard as its chief enemy, the United States.

One view of this situation was given in the summer of 1966, when I dined at the home of the Soviet atomic physicist Dr. Gersh Budker, who then directed the Nuclear Energy Center at Siberia's "science city" near Novosibirsk, 2,500 miles east of Moscow. While enjoying his hospitality in his attractive *dacha,* set in a Siberian birch forest, I discussed with him many aspects of Soviet science, including the state of Russia's military science at the end of World War II. He did not speak English, but as we talked over a bottle of excellent Georgian wine, his wife, Ludmila, quickly translated.

"We didn't have the A-bomb then," he said, "and were afraid of you. We have had a long history of people coming to get us and we were nervous about your big bomb. So we put a lot of troops on the new European border. It was just a show of bodies and force but it covered up for us. We bluffed it."

From a military standpoint, the Germans had left the Russians—and indeed the entire world—behind in rockets, and with the atom bomb the United States had far transcended the Katyusha field-artillery rockets and the world's then best field equipment, the outstanding Soviet Tiger and Hunter tanks, which helped roll back the Germans from Stalingrad.

In the Russians' eyes, they were ironically and most decidedly bested in two areas of military science at the very moment of a long-sought and expensive victory. For the first time in their history, even a million troops in Europe equipped with their best weapons could not guarantee national military security against rockets and thermonuclear weapons.

The Soviets, for all their splendid tanks, subway systems, and universities, were a technological have-not nation. Their sudden inferiority was in the same military area that on the steppes of the Russian heartland they had thought so superb against the vaunted Germans. They had emerged at the moment of victory not as the pre-eminent technological nation of the world, but simply as a power, like Red China, that had an endless source of manpower. No doubt pricking their sensitivities was the knowledge that the worldwide reputation of Tsiolkovsky, who really had given them an early and substantial lead in rocket theory, had been overshadowed by the developmental work in Germany under Oberth. Also, the real prize of Germany, the brainpower that patently had the best-defined ability to forge a truly long-range rocket, had elected to side with the United States. Soviet intelligence would have been blind, indeed, not to realize we had scooped, literally from under their noses, 300 boxcars of rocket equipment and documents westward from Nordhausen and had, subsequently and in great secrecy, spirited the von Braun group to a secret U.S. destination.

The distinguished British author on space affairs, Arthur C. Clarke, once reported a postwar conversation between Stalin and one of his rocket scientists. "This is absolutely intolerable," Stalin is reported to have said. "We defeated the

Nazi; we occupied Berlin and Peenemünde, but the Americans got the rocketeers. How and why was this allowed to happen?"

It is not difficult, in retrospect, for a Westerner to see what appeared to the Soviets as the only logical crash program in the delicate, mutually suspicious, and complex postwar confrontation of East and West.

From the rocket-knowledgeable cadre in Moscow to the highest officer of the Red Army and the Red Air Force, national and psychological insecurities indicated but one logical direction for the United States to take—the merger of the new rocket technology with the new nuclear warheads. To the Soviet policymakers, there was no alternative but to commence two mandatory and related projects: the development of thermonuclear weapons, and the perfection of the nuclear-bomb rocket carrier of sufficient range to reach the North American continent.

Although we now know that these were the two main objectives of Soviet technology, the Soviet internal viewpoint about the importance of the military role of rockets was not unanimous, by any means. While the front pages of the world were preoccupied with man-to-man and leader-to-leader cold-war confrontations at Berlin and elsewhere, the Soviets used the Iron Curtain to shield an intense, high-budget internal objective that was to challenge Soviet technological ingenuity as it had never before been challenged in any so-called peacetime era.

The Rand Corporation's knowledgeable Russian space expert Dr. Fermin Krieger reported to Congress in the late fifties that following World War II Russia "built a large number (probably several thousand) of German V-2 missiles, presumably for fundamental rocket research and upper-atmosphere research, for gaining experience in mass-producing large rocket missiles, and for training rocket-launching crews. They not only improved the V-2 rocket engine . . . they have also developed a super-rocket. . . . These developments indicate that the Russian effort has been more than an extension of previous German work; to all indications it is based on independent thinking and research. This is not surprising since Russia has its share of exceptionally capable technical men . . . in the field of combustion theory and fluid dynamics."

What was the U.S. doing at this time with its Peenemünde treasure trove? Virtually nothing. The Peenemünde group was given a five-year contract and sent to the remote Southwest desert area of Fort Bliss, Texas. I asked a Huntsville German, Walter Weisman, what they did during their first five-year contract.

"We taught each other English," he said, "played chess and occasionally chased empty V-2 fuel tanks which the wind blew across the desert like tumbleweeds. Von Braun wrote a book on a space trip to Mars. All of us were frustrated. We were itching to get on with larger rockets and put an earth satellite into orbit."

It was not until the early fifties that the U.S. Army assigned them to help develop our small (75,000 pounds thrust) Redstone rocket. The eventually reliable Redstone was not fired from Cape Canaveral, however, until August 20, 1953. The Russians, in the meantime, had already accelerated V-2 propellant flow and produced a new rocket that we now know had a thrust slightly greater than our first Redstone. The Russians must have thought a long time before they decided to parade their new rocket in Red Square on May Day—an event that partly awakened a sleeping Pentagon.

The same year that we announced the launching of our first Redstone, a significant hint was dropped by the Russians on what was going on behind their security fences. At a World Peace Council meeting in Vienna on November 27, 1953, the then president of the Soviet Academy of Sciences, A. N. Nesmeyanov, said: "Science has reached a state where it is feasible to send a stratoplane to the moon, to create an artificial satellite of the earth."

Other early evidence was slight. A year later, Tikhonrarov published an encyclopedia article on "Interplanetary Communications," and simultaneously Russia announced the establishment of the K. E. Tsiolkovsky Gold Medal, to be awarded every three years. But few Americans attached any significance to either event. At subsequent international conferences, significantly, the Soviet participants in astronautics were men like L. I. Sedov, of outstanding reputations in science. By now, Russia had established a powerful internal organization that went by the tongue-twisting name of the Interdepartmental Commission for the Coordination and Control of Scientific-Theoretical Work in the Field of Organization and Accomplishment of Interplanetary Communications of the Astronomical Council of the U.S.S.R. Academy of Sciences. From the beginning, in considerable contrast to procedures in the United States, members assigned to this high-priority group, known as the ICIC, were the most outstanding Soviet scientists in a variety of fields—from chemistry to physics and from mathematics to astronomy. Obviously, the U.S.S.R. recognized very early that the science of astronautics embraces nearly every known field of science. When the names of the ICIC members were finally published, some Westerners recognized not only outstanding scientists but also a number of military men, including explosives expert Major General G. I. Pokrovsky and artillery expert Lieutenant General A. A. Blagonravov.

What Russia regarded as the response to its urgent military objective was making progress also. In 1949, certainly with the assistance of efficient espionage, the U.S.S.R. exploded its first atom bomb. Unknown to the West, the Russians were perfecting a means, other than aircraft, for delivering it to a far-distant target.

The Russians can credit former Premier Nikita Khrushchev with energetically pushing the development of the U.S.S.R.'s Strategic Rocket Force, Khrushchev entered the top arena of power, along with Nikolai Bulganin, in 1953, following the death of Stalin. Khrushchev inherited a military system that, like our own, was slow to accept change during the years of so-called peace. Following World War II, the heroes of the Soviet Union were powerful and tradition-rooted Red Army leaders like Marshal Vladimir Sudetz and Red Air Force officers like Chief Air Marshal K. A. Vershinin. Of the two forces, Khrushchev found the Army most resistant to change. He virtually rammed rockets down the throats of the Red Army traditionalists.

There appeared in Russia in 1962 a new book called *Military Strategy*, which subsequently has proved to be the most significant work on this subject published in three decades. In it, the former Chief of the General Staff, Marshal V. D. Sokolovsky, and other Russian military leaders discuss the dramatic shifts in military thinking that began in the late fifties under Khrushchev. The shift, in its broadest terms, was from the Soviet preoccupation with theater land warfare to global strategic warfare involving rockets and nuclear weapons. The phase of this shift most pertinent to rockets began in 1957. By then the Soviet Union had a considerable number of advanced rocket weapons carriers, and Khrushchev, who was now in full command, began his public as well as private debate with the military old guard.

The public portion of this debate was combined with what came to be known as "rocket rattling." Khrushchev began to boast more and more that not only did the Soviet Union have "formidable" weapons but also that "those that are, so to speak, about to appear, are even more perfected, even more formidable." His purpose was threefold: to intimidate the West; to advise the Soviet people that their technological inferiority at the close of World War II had been eliminated; and to break down the insistence of the Soviet military that a large, conventional land army was an entirely adequate defense. Eventually, Khrushchev did drastically reduce the land army (the actual cut was 1,200,000 men), and he defended his action this way: "In our time . . . the defense potential of the country is not determined by the number of our soldiers under arms, by the number of persons in uniform. . . . The defense capability of the country depends, to a decisive extent, on the total firepower and the means of delivery available. . . . The proposed reduction will in no way weaken the firepower of our armed forces, and this is the main point."

The West did not realize it at the time, but this seemingly logical concept of military strategy was by no means unanimous in Russia. Soviet Defense Minister Malinovsky himself made numerous speeches that were polite and cautious but nevertheless openly appeased those of his generals who strongly opposed the "brash" concepts of Khrushchev. The chief points of the dissenting generals were that there was no assurance a future nuclear war would be "swift and short," as Khrushchev believed, and that there was no assurance the West would not attack with conventional land armies. They stressed over and over the "combination of forces" concept. But Khrushchev, eager to budget a large intercontinental ballistic-missile force, stuck by his rockets and largely carried his "new strategy" of a drastic alteration of military emphasis. The "victory" is unequivocally stated in the book *Military Strategy*: "The Strategic Missile forces, which are the main weapons of modern warfare," wrote Marshal Sokolovsky, "will not accommodate their operations to those of the Ground Forces, but vice versa." Once this doctrinal imperative was imposed on Soviet military strategy, the machinery was quickly set in motion for the organization of the elite Strategic Rocket Force under Marshal Nedelin and, later, Marshal Nikolai I. Krylov.

The fact that Khrushchev did emphatically impose his concepts had a significant effect on the dawning space age. As Russia beefed up and mass-produced its ICBMs, the U.S. countered with long-range boosters of its own, and, simultaneously, aerospace scientists of both countries began to base their plans for space exploration on the escalating lifting power of military boosters.

This cause-and-effect response and relationship between military vehicles and exploration has many precedents in history. The ships that first sailed from Europe to the New World either were, or were devised from, military vessels. Similarly, military aircraft was the direct antecedent of the worldwide expansion of air exploration and commercial aviation—from the first passenger and mail carriers to the modern cargo and passenger jets. The pattern of broad technological pioneering following upon the heels of political or military insecurities is an old one.

Nikita Khrushchev's early recognition of the age of rockets and his stimulation of the mass development of large-booster rockets has had a profound and accelerating effect on the rocket age and, ultimately, on space exploration. The American President opposing him at the time, Dwight Eisenhower, was himself a military man, yet his effect on his own nation's strategic policy was more akin to the traditional roles of Marshal Malinovsky and Marshal Zhukov than it was to that of Khrushchev. Eisenhower, who was "soft" on space from the very beginning, found himself in the position of defending an attainment at a time when Khrushchev was embracing and stumping for amalgamated rocket-age-space-age-nuclear-age concepts. The driving force of the U.S. shift to strategic rocket forces was neither Eisenhower nor General Curtis LeMay, who was also defending an attainment. It was, instead, a young, hard-driving Air Force general named Bernard Schriever; it was Schriever and his colleagues who finally convinced Eisenhower, not the other way round.

The pre-Sputnik situations in the U.S. and the Soviet Union provide an interesting contrast: In the U.S., it was neither political leaders nor prominent scientists who urged an affluent and somewhat indifferent citizenship across the initial rampart of space; it was a small corps of hardheaded military men with a clearly defined but limited objective. In Russia, in contrast, the Head of State was himself an energetic trailblazer, backed up by top scientific brains that, from the beginning, sat alongside military counterparts on key policymaking committees. As a result, the pre-Sputnik Soviet position was much more strongly motivated, broader in scope, and more practically organized. It was chiefly these advantages, combined with a rich tradition in rocketry, that enabled the Soviets to more than compensate for the intrinsic handicaps of their more Spartan economy.

4
Before Sputnik

Before Sputnik, the U.S. had three primary sources of evidence of the Soviet evolution of rocket technology. One was the annual parade of military rockets in Red Square. To this, we paid the closest attention; we sent to Moscow as analytical witnesses not only our best experts but also our most skillful photographers. As barometers, however, these parades were far from reliable, since the Russians deployed only what they wished to display for internal or external effect. The second source was the public rocket-rattling and claims of Soviet politicians—claims obviously subject to manipulation under the pressure of current international events. To these, we also gave intense scrutiny—so intense, in fact, that we overlooked the third and most important source of all: what Soviet scientists themselves were saying and writing about their progress. It took, and is taking, the U.S. an incredibly long time to realize that—with certain conspicuous and almost predictable exceptions—the Soviet scientist, like our own, likes to hold himself and his work as remote as possible from the contamination of the more expedient world of politics and propaganda. He is not prone to tell lies that would be readily apparent to his colleagues. The scientist of virtually every nation is normally happiest when he is free to concentrate on his own research and experiments. For the most part, he is congenitally and professionally opposed to and unqualified in the dark arts of exploiting national mentalities and psychoses through a subtly exaggerated or deliberately false arrangement of facts. To this extent, he is somewhat like the front-line soldier who discovers that the farther back he gets through a succession of headquarters, the more obviously corrupt the world becomes. A scientist who searches for truth is severely penalized in his own work if he does not become accustomed to recognizing a fact and dealing with it in terms of reasonable certainty. Such an innovator is suspicious of any pressures brought to bear that in any way alter the facts. The notion that Soviet scientists are any more ethical or unethical than our own in this regard simply assumes that human nature in one part of the world is different from human nature in another. During the late fifties, it came to be the easy fashion in the U.S. to label all reports from the Soviet Union as "propaganda." This included, unfortunately, the speeches and papers of the Soviet scientific fraternity, a source of information that history has proved was largely reliable and remarkably exact.

The exceptions occur, logically, precisely where they usually occur in most countries—at that level where a trained scientist deserts his primary work for whatever identity or prestige he sees in political affiliation and power. Once a scientist is allied with the competitive world of expediency, his concern for truth is often compromised and subordinated. And nine times out of ten, he is an unreliable spokesman. In analyzing the considerable body of evidence that the Soviets gave of their preparations for orbiting an earth satellite and inaugurating the great competitive surge of the space age, it is obvious who told the lies and who did not. Our error—rooted in a certain smugness—was in our assumption that they all told lies, although in most cases it was simply our incredible oversight in not really paying attention to what the Soviets were saying in their long roster of specialized and general professional journals, which scientists use to communicate with one another and with their colleagues abroad.

Immediately after October 4, 1957, that fateful day the Soviets launched *Sputnik 1* and, in so doing, shocked, astounded, and, in some eyes, humiliated the West, a few people looked up what the Russians had said and written preceding the event. To many, here was another surprise. The record, examined in detail, revealed that the Soviets had, not once but numerous times, explained the advanced state of their rocketry and predicted this and other launchings. After November 27, 1953, when A. N. Nesmeyanov, president of the Soviet Academy of Sciences (SAS), told the World Peace Council in Vienna: "Science has reached a state when it is feasible . . . to create an artificial earth satellite of the earth." In the same year, the secretary of the newly formed ICIC reported, "One of the immediate tasks of the Commission is to organize scientific work concerned with construction of an automatic laboratory for scientific research in space. Since it is outside the limits of the atmosphere, such a cosmic laboratory, which will revolve around the Earth as a satellite for a long time, will permit observations of phenomena that are not accessible for investigation under ordinary terrestrial conditions. . . . The creation of the cosmic laboratory will be the first step in solving the problem of interplanetary communications [this term in Russia is synonymous with "space flight"] and will enable our scientists to probe more deeply into the secrets of the universe."

As launch time approached, the statements became even more specific. In August of 1955, for instance, that frequent commuter to international conferences Academician L. I. Sedov said in nearly full candor: "In my opinion, it will be possible to launch an artificial earth satellite within the next two years, and there is a technological possibility of creating artificial satellites of various sizes and weights. From a technical point of view it is possible to create a satellite of larger dimensions than that reported in the newspapers. . . . The realization of the Soviet project can be expected in the comparatively near future. I won't take it upon myself to name the date more precisely."

According to a post-Sputnik U.S. Senate report, in the same year, 1955, Professor K. Stanyukovich virtually drew a verbal picture of *Sputnik 1* He wrote: "A small missile, the creation of which is the least complicated, will probably be launched first. Such a satellite will have the shape of a hollow sphere, the size of a basketball, in which instruments will be placed. The satellite can attain the necessary velocity with the aid of a two-stage rocket which contains the satellite sphere. It is launched at an angle of 45 degrees."

A year later in Barcelona, Spain, an SAS vice-president, Academician I. P. Bardin, said: "The U.S.S.R. intends to launch a satellite by means of which measurements of atmospheric pressure and temperature, as well as observations of cosmic rays, micrometeorites, the geomagnetic field and solar radiation will be conducted. The preparations for launching the satellite are presently being made."

When 1957 dawned, a year that was to be the curtain raiser of the space age, Soviet scientific papers delineated extensive atmospheric studies using sounding rockets (1) to boost large instrument containers that were returned to earth by parachute, (2) to boost instrumented dogs and other animals to heights of well over 50 miles, and (3) to boost dogs and instrument containers through up to 6 Gs of acceleration thrust and up to 2,600 miles per hour. This all should have indicated that the talent-loaded ICIC had not been just writing memos.

In May of 1957, astronomer A. A. Mikhailov clearly delineated the proposed *Sputnik 1*'s orbital period (1½ hours) and outlined the Soviet plan of organizing a network of twenty to twenty-five sky observers to plot *Sputnik*'s visible course across the heavens. Then, on June 9, 1957, Academician Nesmeyanov laid a clear prediction on the line: "Soon, literally within the next months, our planet Earth will acquire another satellite."

Speeches, scientific articles, and Soviet press reports of equal frankness continued right up until October 1. In late September, the Russian postal department even issued a special Tsiolkovsky stamp picturing a rocket and a satellite. Then, just four days before the launch of the world's first artificial satellite, for anyone who still doubted Soviet intentions, the Russians provided the most unmistakable evidence of all; they announced that their first satellite would broadcast from space on the 20- and 40-megacycle bands. This conclusive evidence of an imminent and historic satellite launch was monitored and promptly passed on to President Eisenhower. His reaction was to do nothing to advise the American people of an event that would ultimately prove more shocking to them and their way of life than anything else that had happened abroad in this century—short of the outbreak of war.

During this pre-Sputnik period, as United States prosperity seemed miraculously to renew and extend itself, we, first of all, simply failed to appraise or admit the technological coming of age of the Soviet Union. We had been given every opportunity to excel in astronautics. Instead, we fell seriously behind. Part of the new apathy of the proud and pioneering American people was manifested in our refusal to admit that another powerful and increasingly modern nation was rising like a Phoenix from the fire nest of World War II. Moreover, the stirrings of an emerging national identity were not, as many conveniently believed, confined strictly to space. The Soviets, long behind Western standards of living, were reaching toward a number of the material perquisites that many Americans felt to be peculiarly their own. In 1953, for instance, Russia opened its huge Gum department store in Moscow. By 1956, Mikoyan's promise of a production of 1,000,000 television sets was exceeded. There was a general decrease in terror and the more brutal forms of censorship that had long been Soviet practice. In 1956, Boris Pasternak published his great critical novel *Doctor Zhivago*. The state criticized and ostracized him, but he remained alive and continued to write. The Soviet theater and cinema rebelled against standardization of style, and the entire country went on an incredibly energetic sports binge. The Russians were still not confident enough to host the Olympic games, although there was no lack of confidence or accomplishment in their athletes. It was, in general, a period characterized by a long-promised but often-frustrated national awakening—an awakening that quickly adopted the dramatically successful Sputniks as its symbol.

Russia's "de-Stalinized fifties," as Soviet historian Basil Dmytryshyn labeled them, were the precursor of what he called "the cosmic sixties." Our own fifties are too recently upon us to permit overall review, but what we were doing, or not doing, in astronautics is, of course, germane.

In 1954, the U.S. began plans for Project Orbiter, with a goal of studying the possibilities of launching a scientific earth satellite. It was a low-budget, low-key project and consequently received a low-key response from the public. Then, a year later, the White House did spark some public interest by canceling Project Orbiter and announcing Project Vanguard, whose orginal objective was to launch a very small artificial earth satellite by 1958—as part of the International Geophysical Year that was to begin in 1957. Even though we then had under development the more powerful IRBM and ICBM boosters, it was decided that our first satellite needed a tailor-made booster. Since the satellite was to be a small one, the booster, manufactured by the Martin Company, was scaled down to the minimum thrust necessary to obtain orbital speed for a light payload. Thus, the slender Vanguard rocket arrived at Cape Canaveral about the time the more powerful Douglas Thors and Chrysler Jupiters were engaged in a neck-and-neck race for selection by the Pentagon as a mass-production weapon. The three services were fighting over which would get what jurisdiction over which rockets and space missions, while the creative brains of U.S. rocketmen (von Braun and his group, now augmented by a number of Americans), had been repeatedly frustrated despite their impassioned arguments for permission for the go-ahead and relatively minor funding to launch an earth satellite.

It was in this atmosphere of competitive bickering—and lobbying, among both the services and giant corporations—that Project Vanguard was given over to the jurisdiction of the U.S. Navy, in the absence of any major Navy role prior to Polaris in strategic rocket development.

And to confound the confusion—or to conceal a general leadership astigmatism—Project Vanguard, which Eisenhower had bragged to the world was "purely scientific," was unaccountably classified. This was particularly frustrating to those who were excited about Vanguard's mission and were trying to report its concepts and progress to our newspapers and magazines. Inquiries at Cape Canaveral were sometimes imperiously evaded under the artificial cloak of classification, and those of us who specialized in space affairs soon learned to circumvent official silence by some privately developed devices of our own. One reporter friend of mine furtively climbed a ship's mast in Port Canaveral harbor and shivered in the crow's nest for six hours while he studied the Vanguard pad with binoculars. Others attempting to obtain some knowledge of the program had their license numbers taken down by the police. In late 1957, Cape Canaveral had not experienced the waves of penalizing strikes, but the general mood, taking its cue from the miasmic state of political, military, and industrial leadership, was not one of either single-mindedness of purpose or dedication to a manifest national objective.

When October came around, the chief hangout for off-duty missilemen was the famous cocktail lounge of the Starlite Motel (which later burned), and the chief topic of conversation around the television sets was whether Casey Stengel's New York Yankees could take their third World Series in a row, this time against Milwaukee.

5
October 4, 1957

IN THE long chronicle of human history, it would be difficult to find a single "peacetime" event that generated such diverse and powerful reactions as occurred in the Soviet Union on October 4, 1957. The launch of *Sputnik 1*, the world's first scientific payload to achieve an earth orbit, exploded with all the impact of a cataclysm—so ostensibly sudden was its dramatic appearance and so consummate was its feat of scientific daring. In rising above the atmosphere and—thanks to the laws of physics—staying there, *Sputnik 1* was, first of all, the auspicious curtain raiser of the age of space. When it thundered up from its arcane nest of fire, it became the pathfinder for the untold tons of strange new vehicles and the curious explorers who were to follow its wake into an entirely new dimension of human inquiry.

It created an international impact that was virtually measureless at the time and whose assessment will still be attempted in those eons to come that will have the advantage of perspectives. Never before had a nation widely believed to be second rate in education and technology so drastically demonstrated in a single conspicuous stroke that it had forged a revolutionary and potent new instrument unmatched anywhere on the face of the globe. Its launch immediately precipitated in the United States one of the most intensive soul-searchings and re-evaluations we have ever undergone. In three areas at once—military, academic, and scientific—we pondered our assets and took stock of our real or suspected deficiencies. Although eventually the en-

Konstantin Edwardovich Tsiolkovsky (1857-1935), the "father of Soviet rocketry", in a photograph taken in 1934. Inventor and prophet of space travel, he designed early models of dirigibles, airplanes, a wind tunnel, and, prior to Robert Goddard of America and Herman Oberth of Germany, contributed to early rocket research and theory. The inscription above a rocket design of 1903 reads: "The impossible today will become the possible tomorrow".

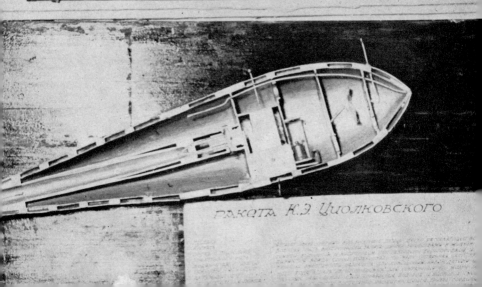

(*Above*) Laika, a black and white mongrel fox terrier, was the first living creature to orbit the earth. Launched on November 3, 1957, aboard *Sputnik 2*, Laika radioed back data on her physiological reactions to the stress of space travel for seven days, providing invaluable information in the research on manned space flight. (*Below*) Polkan and Gray, a dog and a rabbit, in space suits preparatory to a flight into space in the 1950s as part of the early Soviet use of animals to investigate the health hazards of space travel.

tire fabric of U.S. education was altered in reaction, the first effect of Sputnik was one of widespread consternation. Patriotically and psychologically, the U.S. was unprepared. Our postwar mood still reflected the pride of the victor and the mixed and expedient blessings of having been the first to harness the mighty atom. Now the image of an unsurpassed technological virtuosity and a prosperity-nourished complacency had been precipitously shattered.

But our national pride was yet to receive two more related blows. One month later the Russians orbited the 1,120-pound *Sputnik 2* with a live dog aboard. Then they capped this in early 1958 by orbiting 2,925 pounds—more than twice that weight—in *Sputnik 3*. In the meantime, the several hundred of us who had assembled on the sands of Cocoa Beach had the unpleasant task of reporting to the world the humiliating explosion of Vanguard on its pad. Its tiny, 3.2-pound pretender to the sacrosanctity of space, which Khrushchev immediately dubbed a "grapefruit," lay where it had been blown by the explosion—on the Cape Canaveral palmetto flats. It was plaintively broadcasting its frustrated signal not from the serenity of the heavens but from the ground. As the entire world knew, the slender, delicate, and underpowered Vanguard rocket had toppled into its own ball of fire. Amid the twisted wreckage, flames licked at the charred shell of a national aspiration. Vanguard's inferno reflected the Sputnik achievement in a red-stained mirror.

"A few values of very great importance were at stake, pride above all," said the University of Iowa's social scientist Vernon Van Dyke. "It is exaggerating considerably to compare the situation to the one created by Pearl Harbor, but still the analogy is suggestive."

In Huntsville, Alabama, Wernher von Braun said: "The reaction to these events has been profound. They triggered a period of self-appraisal rarely equaled in modern times. Overnight, it became popular to question the bulwarks of our society; our public educational system, our industrial strength, international policy, defense strategy and forces, the capability of our science and technology. Even the moral fiber of our people came under searching examination."

In Europe, the London *Times* wrote of "the demon of inferiority which, since October 4, 1957 . . . has disturbed American well-being."

When Americans looked, after *Sputnik 1*, to the highest level of their political leadership, they found a strangely detached and mollifying tone. President Eisenhower's words on this occasion have often—and usually derisively—been quoted: "Now, as far as the satellite itself is concerned, that does not raise my apprehensions, not one iota. I see nothing at this moment, at this stage of development, that is significant in that development as far as security is concerned, except, as I pointed out, it does very definitely prove the possession by the Russian scientists of a very powerful thrust in their rocketry, and that is important."

In what can only be presumed to be his ignorance, the President gave powerful fuel to another of the regrettable myths that was to grow across the land and becloud the issue. Five days after the launch of *Sputnik 1*, Eisenhower made his unfortunate suggestion that the reason the Russians were first in space was that they had "captured all those German scientists at Peenemünde."

It was not until years later, on December 1, 1966, that Army General James M. Gavin revealed some of the U.S. shortsightedness when, from the invulnerability of retirement, he told the American Institute of Aeronautics and Astronautics that a year before Sputnik he had received a written order forbidding him to proceed with the development of an earth satellite.

"We believed in 1956," said General Gavin, "that we had the capability of orbiting an earth satellite. . . . On the basis of this I made several entreaties to the Department of Defense seeking authority to launch a satellite, and shortly thereafter I was given a written order forbidding me to do so. This admonition was passed on to von Braun."

As a matter of fact, Gavin was discussing this remarkable prohibition with von Braun in Huntsville, Alabama, the day *Sputnik 1* went into orbit. "We returned next day to Washington," Gavin said, "and the Washington headlines on the morning of October 5 reflected the hysteria of the hour."

A few days later, he reported, a member of the White House staff referred to *Sputnik 1* as a "silly bauble."

"The point I want to make now," Gavin told the AIAA, "is that this degree of ignorance of what the Soviets were capable of doing with no apparent capability on our part was totally unacceptable to the American people, as the White House soon found out."

The situation in the Soviet Union was one of joyous fulfillment. In their arduous and competitive struggle with the West, involving all facets of national strength, the Russians had suddenly found a new and endemic outlet that triggered both national pride and international recognition. The incomplete and in some ways disturbing victory of World War II had been partly redeemed in a bloodless battleground that had the additional and embellishing sanctity of science. The cold war was not an Olympic contest; it possessed none of the restraints of sportsmanship. Khrushchev, who often relished the sensation of having a local audience with him, openly appealed to the puissant elements of international one-upmanship. This jocular and derisive approach—in which the Soviet people joined—not only carried the day over the diehard, theater-warfare Red Army traditionalists, some of whom had witnessed the ICBM booster that lofted Sputnik, but also gave him a brand-new instrument of national policy he was quick to exploit around the globe. His victory was all the sweeter because the West had disbelieved and depreciated much of his blustery rocket-rattling and had, in fact, been taken far more off guard than he —or any other Russian—had supposed possible.

Before launching Sputnik, the Russian scientific leadership had been under the impression they had rather candidly telegraphed their punch. As a matter of fact, they had; but, as we have seen, no one in the West paid any attention. As in the case of Pearl Harbor, one prescient U.S. intelligence report got to the White House just prior to launch, but it had lodged in the same niche of obscurity as far as the public was concerned—that had contained the ill-fated missive that telegraphed the battle stations of the Imperial Japanese Navy in 1941. Even without benefit of intelligence, the presaging message was clear. In August of 1957, the Russian people learned that the Soviet Union had completed a series of ICBM tests that sent payloads out of the atmosphere for a 6,000-mile ride to the target. The next month, they learned Russia was planning to place in orbit two different types of earth satellites, as Radio Moscow repeatedly predicted that "in the near future" the Soviet Union would launch a satellite. Then, just three days before launch, it announced the *Sputnik 1* radio-transmission frequencies, and anticipation ran high.

Meanwhile, convinced that they had telegraphed their intentions to a listening world, Soviet scientists, artillerymen, and airmen groomed a powerful three-stage rocket on its pad near Tyuratam, located in the heart of the old Genghis Khan empire just east of the Aral Sea. Our then-secret tracking station in Turkey had established that Soviet geophysical rockets had often been launched from Kapustin Yar, about one hour's train ride south of Stalingrad (now Volgograd). But KY, as it was designated on U.S. intelligence maps, was located in the low-lying and often misty Volga River basin, and the prevalent moisture was injurious to sensitive instruments and equipment. KY security was also, in Russian terms, almost as bad as the "goldfish bowl" of our own Cape Canaveral; KY rockets could be seen rising, for instance, from Soviet excursion boats and speedy hydrofoils that plied the Volga between Stalingrad and Astrakhan on the Caspian Sea. Also, from KY, the preferred satellite-orbiting inclination—toward the southeast—passed over several populated areas where the abortion of faulty rockets could risk human life.

Their long-term needs were for a drier climate in a less populated area whose remoteness provided an increased element of security combined with rail and water access to the rocket plants that were largely concentrated in more industrialized western Russia.

The Tyuratam launch complex (sometimes called Baikonur by the Russians) met all these conditions. Although located near the north-flowing Syr Parya, the surrounding countryside was semiarid desert. Preferred launch inclinations were over vast, uninhabited desert areas, and from a security standpoint, Tyuratam (or TY, as designated on U.S. intelligence maps) was more than twice the distance of KY from our known tracking installation just over the Turkish border. TY, in addition, was adjacent to the wide-gauge railroad running south from the industrial north to Tashkent in mountainous southern Russia.

The actual site selected at Tyuratam was a broad, nearly perfectly flat valley that, thanks to rudimentary irrigation, supported a variety of grasses and a few trees, including poplars. The initial installation was a cluster of low white buildings and concrete blockhouses not too unlike those at our own desert base at White Sands. One central building had a long, second-story, old-fashioned-looking wooden balcony from which observers with binoculars could get a clear view of the *Sputnik 1* rocket as it fumed and smoked, a monumental, candle-like rocket standing erect and entirely alone on a flat sea of grass.

On the morning of October 4, 1957, the multiple-stage rocket must have presented an imposing sight to those privileged to witness its dramatic premier. During the many years that the Russians neither said nor wrote anything about the external configuration of the Sputnik booster, U.S. specialists worked overtime trying to determine its characteristics. A remarkably specific report, based on the assumption that its main stage was the same ICBM that had just passed its 6,000-mile-range test, was made available to the author in the early sixties. This report maintained that the overall thrust was in the neighborhood of half a million pounds. The center engine, an enlarged and improved V-2 engine of about 220,000 pounds thrust was surrounded by four smaller engines that provided a thrust beyond that needed to orbit the satellite. In 1967, the Russians finally released a picture which, if correctly captioned, partially validates the earlier U.S. report. The rocket is seen to have a somewhat stubby main booster surrounded by four Vostok-like rocket pods or cones tapering out to an array of nozzles at the rocket's base.

There was nothing particularly exotic about its complex

internal plumbing. The arrangement of fuel tanks, turbo pumps, corrosion-resistant and fuel-cooled combustion chambers and flared nozzles was essentially the same as that—based on pilot V-2 principles—used in the United States.

On top of this ICBM, which alone had the power if not the cut-off precision to orbit the 184-pound Sputnik payload, was the rocket that was to give Sputnik its final, exactly calculated and aimed orbital boost of about 1,700 miles per hour. At the rocket tip, encased in a detachable fairing, was the shiny ball of Sputnik (which means "satellite" in Russian). When I inspected a *Sputnik 1* replica in Russia, it seemed a far less significant sphere than one would suppose from its role in history. At 23 inches in diameter, it was the size of a beach ball. Its frail wings were four folded antennas —two of them about eight feet long, the other two a foot longer.

Inside the 184-pound Sputnik ball was a reliable radio transmitter set to send its now famous beep-beep on two readily accessible frequencies. Also packed inside were small instruments to gather and radio information on density, temperature, cosmic rays, and micrometeoroid activity, as Russian scientists had foretold.

Inexplicably, the Russians, who in the 1960s became frank about many scientific aspects of their space program, have never published a description of whatever suspense, awareness of history, or emotional content characterized their preparations at the launch pad. Their dedication and often demonstrated compulsion to be first certainly contributed considerable pressure. They well knew of the Vanguard program that the U.S. had announced in 1955, along with its plan to launch a series of small satellites during the International Geophysical Year. And they were assuredly aware that Vanguard was already at Cape Canaveral and potentially, at least, could orbit ahead of *Sputnik 1*. In fact, as a journalist I had reported the fact that as early as December of 1956, a then unaccountably classified, pre-Vanguard Viking rocket called TV-Zero had carried a small, round test satellite that had improvised antennas made from metallic measuring tape. This was a ballistic shot only—orbit of the test satellite was not attempted—but the resulting exclusive story went round the world, and the Russians were probably even more aware by other means that, in our fashion, we were gearing up for an imminent satellite launch.

The man in charge of the historic launch was the *Sputnik 1* technical director, a man who has since become a Russian legend and who, along with such men as Wernher von Braun, Bernard Schriever, Kurt Debus, Kraft Ericke, and the astronauts of both East and West, belongs in the select handful of men who played major roles in the conquest of space. The Russians themselves never revealed his name while he was alive. It was only after two months of intensive investigation that this writer finally felt certain enough of his identity to name him in print in 1966.* His name was Sergei Pavlovich Korolev, and we shall be hearing more about this remarkable aerospace engineer as this book progresses. At the time of the Sputnik launch, he was 49 and, according to Yuri Gagarin, "a stout man of medium stature. He has a receding hairline, deep-set brown eyes, a small nose, and small, rounded hands." He was described by others as "cheerful and witty," although somewhat stern in appearance, "even a bit gloomy, but when he starts to talk he exhibits warmth, charm and nobility." During the years when he was identified only as the "chief spacecraft designer" or "chief constructor," colleagues often spoke of him with the unabashed admiration reserved for those few men who lead by the force and sincerity of their character rather than by whatever rank they go by.

"The chief designer," wrote Professor Petrov in *Pravda Ukrainy* in 1963, "built the carrier rockets as well. The clean outline of this silver rocket was born in his head. He is the prime developer. He placed instruments in rockets and later made a place for man. His word is the final one on the launch pad."

---

* *Fortune*: "The Russians Mean to Win the Space Race," by William Shelton, February, 1966. He was identified in print just eight days before his death from cancer, after which the Soviet Union, for the first time, publicly recognized him during his hero's burial in Red Square.

Korolev was originally a pilot and had risen to a lieutenant general of aviation, but when he transferred to the Soviet rocket program, he chose to wear civilian clothes. He sometimes appeared in a white coat and straw hat. From all that responsible Soviets, from cosmonauts to engineers, have said and written about him, he evidently possessed the highest qualities of leadership, superb talents as a designer, and the kind of firm but benign personality that inspires an unusually high order of hero worship. "He listens attentively," one close observer noted, "then when he speaks, his clear and accurate instructions are carried out to the letter." A Russian rocketeer named Peskov once wrote: "You talk awhile with him and you feel yourself charged up for the flight, especially on the eve of a launching . . . I always waited for his encouraging words." He apparently commanded a certain awe by his presence alone. An engineer who worked closely with him wrote that whenever Korolev held a technical conference in the assembly shop boasting was frowned upon. The discussions were very short, terse, and precise. As a rule, no one volunteered his opinion. Korolev could be sharp and abrupt, but he was never spiteful.

It is not yet known what Korolev may have said before and after the *Sputnik 1* launch, but it is now known that he was chiefly responsible, along with his colleague, Valentin Petrovich Glushko, then also 49, and another leading theoretician and engine designer, M. K. Tikhonrarov. Shortly after Sputnik, both Korolev and Glushko were elected to the august Soviet Academy of Sciences, but their specific duties were purposely never delineated.

After Korolev's death on January 14, 1966, and his Kremlin burial, attended by the cream of Soviet political and scientific leaders, *Pravda* paid him an unusually high tribute, with the certain knowledge that it would never, for political reasons, have to be retracted—as somewhat similar encomiums to Stalin and Khrushchev have had to be later amended. "With the passing of S. P. Korolev," wrote *Pravda*, "our country and the world's scientific community have lost one of the most outstanding scientists in the area of space-rocket technology. Korolev was the designer of the first artificial earth satellites and spaceships which marked the beginning of the era of cosmic conquest by mankind. . . . The inexhaustible energy and talent of this scientist-explorer, his amazing engineering intuition and his great creative courage, Korolev combined with brilliant organizational capabilities and high moral qualities."

Pravda indicated that Korolev was born the son of a teacher in Zhitomir, graduated from the Bauman Technological Institute in Moscow in 1930, the same year he became a pilot. After meeting Tsiolkovsky, Korolev became fascinated with space-rocket technology, but not entirely to the exclusion of airplanes and gliders, some of which he also designed. He was one of the early members of GIRD, which was formed in Moscow in 1933.

The Soviets were quick to recognize and reward his unique contribution with the title of Hero of Socialist Labor, the Lenin Prize, and others of the highest orders and honors of the Soviet Union. He was unknown abroad, of course, until his death, but upon that occasion *The New York Times*, sensing his significance as much from the caliber of the people who attended his funeral as from any intimate knowledge of his work, editorialized: "Nikita Khrushchev utilized each Soviet space feat as a means of enhancing his own image, and his successors have done the same. None of these politicians wanted to share the spotlight with a nonpolitical technician who did not even join the Communist Party until 1953.

"Korolev's rockets were powerful enough to send men into orbit and to put cameras into position to photograph the back side of the moon. But they were too weak to break the chains of secrecy that denied him, while he lived, the world applause he deserved."

Although it is not yet ascertainable precisely what Korolev did on and preceding the morning of October 4, 1957, we do know that he was there at the launching.

Immediately following Korolev's death and the Soviet's public identity verification, I cabled a journalist friend in Moscow to send all available pictures of Korolev. I already had in my possession a rare picture of an unidentified stocky man standing alone at a rectangular blockhouse window watching *Sputnik 1* rising in the distance. When the Moscow pictures arrived, they matched the mysterious pensive figure by the concrete window. It was Korolev, and his isolated stance by the periscope was tacit testimony to his responsible role and unique rank.

Sputnik 1 belonged, as much as to any other man, to Sergei Korolev. As the gleaming sphere went securely into an orbit of 140 miles perigee and 587 miles apogee, Radio Moscow broke the astounding news to the world and ticked off the then almost unbelievable schedule of appearances over such cities as Yakutsk, Prague, Rangoon, Bombay, Damascus, Paris, and Rome. Its weight, altitude, contents, purpose, and orbital inclination (65°) were all announced but seldom believed, or, if finally believed as our own tracking facilities belatedly tried to lock on the orbital path, often depreciated as "propaganda." This pattern of conveniently designating Russian space accomplishments as "propaganda" thus got an early start—thanks to both Khrushchev's flamboyant boasts and the native psychology of a large, and even influential, segment of our society that refused to believe that our vaunted technology could be bested by the "clumsy-handed Russians" who had "got all those German scientists at the end of the war."

One person who had no such myopic view of the Russian accomplishment was Dr. Henry L. Richter, Jr., then with California's Jet Propulsion Laboratory. Immediately after the Soviets announced *Sputnik 1*, Dr. Richter, together with the San Gabriel Valley Amateur Radio Club, improvised a tracking station in a small room lent by the Los Angeles County Sheriff's Department. Using equipment donated by local industry, Dr. Richter hastily designed a tracking antenna system known as "Microlock," which was adjusted to 20 and 40 megacycles. Nearly a year later, on a trip to Moscow, he was amazed to discover that his were the most useful recordings made of the first three Russian satellites. A Russian electronics specialist, Andrei Chudakov, asked Richter if it would be possible for the Soviet Academy of Sciences to have a copy of the tapes. Richter agreed, provided the Russians furnished him with the code to the signals, which they promptly did. It appears that the Soviets had had trouble with their own data-recording equipment. Thanks to Dr. Richter's improvisation, they later got a good record of the *Sputnik 1* signals.

Sputnik 1 circled the globe once every 96 minutes and sailed serene and unchallenged until its batteries died on November 14. Then, no longer broadcasting scientific data, it coasted for nine more weeks until its orbit decayed and it perished in a streak of incandescent flame against the harsh friction of the earth's atmosphere.

6

The Head Start

AFTER *Sputnik 1* went up, Korolev, his rocket specialists, and aerospace biologists Dr. Oleg Gazenko and Dr. Vladimir Yazdovski concentrated on launching a biosatellite as soon as possible. Other work proceeded on even heavier satellites. In addition, they had in the wings a modified and more powerful ICBM to be used as the booster for their heavier payloads. So nearly ready were these new rocket-satellite combinations and so complete was their instrumentation that it is reasonable to suppose that the comparatively simple *Sputnik 1* was redundant to the scientific portion of their objectives in every sense except booster guidance and control and their 21-station tracking-range checkout. From the standpoint of prestige and historical significance, however, there is nothing superfluous about an object of whatever size and sophistication that gets there first.

In the Soviet jubilation following *Sputnik 1*, even the normally ebullient and garrulous Khrushchev was partly held in check by his certain knowledge of what was soon to come. The primary shock of *Sputnik 1* was the engineering legerdemain of achieving an intended earth orbit on what, so far as the West knew, was the very first try. And 184 pounds, compared to what Russian leaders knew was Vanguard's intended initial payload, 3.2 pounds, was vastly superior. But while Vanguard could not have lofted the *Sputnik 1* weight, our Atlas, Jupiter, and Thor eventually could, according to the authoritative counsel of those who belatedly briefed the now-attentive ears at the policymaking level in

THE HEAD START

Washington. None of these U.S. rockets was ready, but they could be readied within a reasonable time. It was the precision, not the suspected size, as the *Sputnik 1* booster train, that generated the most powerful shock wave. To some people, Khrushchev's confident boasts of truly formidable and transcendent hardware still seemed to have been an exaggeration designed for effect.

In his book, *The First Four Stages*, published in Russia in the spring of 1968, an engineer associate of Korolev, Alekse Ivanov colorfully described preparations for the next launch. "The carrying out of the objective proceeded at full speed. The bays of the assembly shop resounded with the barking of dogs. Days and nights fled swiftly; men forgot about rest—resolving problems, arguing frightfully, abusing each other and accusing each other of God knows what! In the plant there appeared, now here, now there, Vladimir Ivanovitch Yazdovsky and Oleg Georgievich Gazenko, organizers and inspirers of the biological experiments. Constantly the tall figure of Aleksandre Ivanovich Ephremov rambled about. The stocky Serge Nikolaevich Vernor, with disheveled hair, sometimes joined him. These were the two cosmic ray investigators. But over all this seeming confusion was the iron hand and will of Serge Pavlovich Korolev."

On November 3, one month after *Sputnik 1*, Korolev who was informally known as "S.P.," and his colleagues launched a 1,120-pound payload that contained not only more extensive data-gathering instruments but also an aluminum environmental canister that cocooned a live dog named Laika just beneath a complex, cone-shaped array of temporary life-support equipment and data-gathering instruments. For those doubters of the Soviet achievement—and there was still an abundance of doubters—the empty *Sputnik 2* rocket casing, which remained in orbit still attached to the payload, provided man-in-the-street proof of its existence; the slowly tumbling booster could be clearly seen as it slowly traced a path across the skies above the United States and in many other parts of the world.

Sputnik 2's added weight was not the only unexpected attribute; in lofting a live and instrumented animal into orbit, the

Soviets again demonstrated the unassailable logic of their chronology. Our own program eventually called for biomedical payloads, including two mice named Mia and Wickie, plus yeast, onions, and sea-urchin eggs, but it was not until May of 1959 that we were able to send two squirrel monkeys named Able and Baker into space. Now, on the heels of *Sputnik 1*, the Russians were already receiving detailed biomedical data from a black-and-white mongrel fox terrier. They were thus the first to tell the world that environmentally controlled life could survive the combination of acceleration forces, weightlessness, and cosmic rays that were neither filtered nor absorbed by our atmosphere. The fact, as announced by the Russians, that life could sustain itself, at least briefly, in the space environment was a revelation that wiped clean one widely speculated barrier that many still held would ultimately confine man's destiny to his native planet. Few reputable scientists then took for granted that the space environment was inimical to life.

As we were to find out gradually, supporting life in the alternately frigid and baking vacuum of space was a highly complex undertaking. And the Soviets could not have done it in *Sputnik 2* without a long developmental period of trial and error. Soviet rocketmen who, as we have seen, were thoroughly versed in the more mundane aspects of their new technology were also devoted students of space-science fiction—especially the works of the Englishman H. G. Wells and the Frenchman Jules Verne. In the Russian translation of Jules Verne's *From the Earth to the Moon*, published in the 1860s, there appears a vintage illustration that shows chickens and dogs—as well as men and telescopes—floating in a weightless state. It is a curious coincidence that the 14-pound dog Laika resembles in size and markings the dog depicted in the Jules Verne illustration. While American scientists tend to favor such primates as monkeys, chimpanzees, and baboons for man-related medical experiments, Russians traditionally favor dogs; famed Russian physiologist Ivan Pavlov even erected a stone monument to the creatures that helped him develop his important theories of conditioned and unconditioned reflex.

At the time of Laika's flight, the secretary-general of the Soviet Academy of Medical Sciences was the blue-eyed scientist Dr. N. M. Parin. In Moscow in 1966, I talked with Parin for over an hour. In reply to my question as to why Russians preferred a dog to a chimpanzee for space experiments, Parin smiled broadly. "The Russian dog," he said, "has long been a great friend of science. We have amassed much data on our four footed friends. Their blood circulation and respiration are close to man's. And they are very patient and durable under long experiments."

Academician Parin and other Russian scientists had familiarized themselves with the surprisingly long history of using animals as initial substitutes for men in adventures above the earth. Even Jules Verne knew of the first balloon flight in which the Montgolfier brothers of France in 1783 sent up a rooster, a sheep, and a duck. Tsiolkovsky was familiar with the first animal-carrying rocket flight in 1806 of the inventor Rugieri, and theoretically, at least, had formulated the delicate adjustments and arrangements required to propel animals into space with rockets.

In the late forties, Soviet physiologists, biologists, and aeromedical teams began detailed preparations for sending animals aloft with rockets. Their first experiments were, of course, on the ground, testing responses to acceleration, vibrations, noise, and abnormal pressures. Soviet centrifuge tests showed that animals are least affected when acceleration forces are applied transversely, that is, at right angles to the backbone. They worked out animal space suits and canisters and perfected them in altitude chambers of the Soviet Academy of Medical Sciences in Moscow. The initial rocket launchings of mice, frogs, and dogs in the mid-1950s took place just outside Moscow and at Kapustin Yar on the banks of the Volga River. The first flight of an encapsulated dog was to an altitude of 49 miles. The second flight went to an altitude of 21 miles. Later flights, each bearing aloft two dogs, went to altitudes of 130 miles and, finally, to 279 miles, or 139 miles higher than Laika's point of insertion into orbit. All the earlier animal canisters were designed to be returned to earth by parachute.

The Russians thus discovered prior to *Sputnik 2* that animal weight during a rocket launch increases no more than 5 grams. Their early biomedical instrumentation was highly sophisticated, and established even before Laika that weightlessness, for the five to ten minutes tested, caused no essential electrocardiogram irregularities or blood-pressure deviations and had virtually no effect on respiration. Motion-picture cameras recorded the strange, whirling antics of unrestricted white mice caught in the frictionless crucible of the complete weightless state. Other experiments used animals in various states of narcosis in order to test, as the Soviets put it, the reaction of animals as "living machines."

The paramount pre-Sputnik finding of Soviet bioastronautics specialists Yuganov, Afanasyev, and Kasyan was that animals, including man, could survive in weightless space. An interesting subfinding, which fortunately did not affect the above conclusion, was that dogs tossed their heads going into weightlessness, due to a discrepancy between the neck muscles and the weight of the head; the tension of the dogs' antigravitational muscles had a tendency to linger. Some disruptions of motion and spatial disorientation were also noted in mice, which were contained and photographed in a clear glass box about the size of a hatbox. Russian motion films show the unhampered mice spinning and twisting frantically for up to three minutes in antic and uncoordinated response.

Despite scores of such experiments with ballistic rockets, the Russians still had no data on the effect of prolonged weightlessness on living organisms. In planning the *Sputnik 2* experiment, they made, first of all, no provision for bringing Laika back alive; the black-and-white mongrel was destined to become one of the first of the over 100 animals to be sacrificed by both East and West to the cause of early space-flight investigations. Laika would be supplied with food, however, in order to sustain life long enough to receive radioed biomedical data of some significance. Within these parameters, the Russians proceeded with a knowledge born of extensive experience in instrumenting animals.

First, Laika and a number of possible substitute dogs were

THE HEAD START

given detailed training both to accustom them to equipment and to record, for comparative purposes, their normal and individual earthbound reactions and medical indices.

Laika's sealed cylindrical container was designed to keep its inhabitant alive for one week. A small electric fan recirculated air after its carbon-dioxide and water-vapor content had been chemically scrubbed out and it had been reconstituted with oxygen. Temperature was controlled by a special heat-absorbent screen perfected on the preceding ballistic flights. Body wastes were contained in a rubber reservoir attached to Laika's pelvic area. A special dispenser provided a balanced diet of food and water in the form of gelatin.

Just prior to flight, silver electrodes previously affixed just under Laika's skin were firmly attached to leads running to a miniaturized electrocardiogram. Monitoring the animal's movements was another device called a potentiometric sensor. All instruments were hooked up to radio transmitters. Finally, Laika was affixed to small chains that somewhat restricted her movements but permitted her to stand, lie down, or sit.

Parin and other monitors closely watched Laika's heart and respiration rates as *Sputnik 2* roared up from Tyuratam. They saw both pulse and respiration rates rise during liftoff. Then, as G forces became even higher, they saw the pulse rate begin to descend. Finally, as Laika became the first living satellite of earth, they were gratified to observe near-normal physiological functions.

Before Laika's oxygen expired seven days after launch, those Americans as well as those Russians who had feared that space would profoundly disrupt the basic physiological functions were reassured. Laika's data was not complete by any means, but she had established, as the Russians stated it, that "beyond doubt animals tolerate satisfactorily conditions closely similar to those of space travel." As Alekse Ivanov later wrote, "Laika's flight made it possible to speak more boldly and concretely of the possibility of cosmic journeys of men."

One widespread and initial U.S. reaction was that both *Sputnik 2*, carrying Laika and instruments, and *Sputnik 1* carrying only instruments, were primarily propaganda stunts. The cliché "space spectacular" came into our language along with its connotation that the Russians were putting on a show purely for its theatrical effect on world opinion.

Those U.S. scientists who took the trouble to read translations of Soviet reports and conclusions based on the first two flights, however, knew differently; the world's first two artificial satellites were packed with scientific data-gathering instruments that were imaginatively designed to glean maximum knowledge of the space environment. The world's entire scientific fraternity was eager to know more about the ionosphere and its structure, the influence of the sun's radiation and cosmic rays on the earth's atmosphere, and the nature, density, and temperature of the earth's electromagnetic field. It was precisely these areas that the Sputnik instrumentation was designed to investigate. It was also obvious to some perceptive scientists that in selecting an orbiting inclination of 65° from the equator, the Soviets assured themselves of taking their vital measurements over a wider range of latitudes than would have been possible had they elected the much simpler task of orbiting their instruments around the equator.

In addition, they had made elaborate and scientifically logical plans for tracking the satellites and accurately observing such things as orbital decay due to atmospheric drag forces that gradually and measurably increased as the satellites' elliptical orbits brought them closer to earth. Their tracking and measurements employed both special photographic theodolites that combined a surveying instrument with long-range lenses, and "radiotechnical" devices that obtained accurate altitudes with special radio receivers and recording chronographs. These later fully utilized the Doppler effect—now very common in space-flight instrumentation. This effect, based on a perceptible change in frequency as a satellite approaches or departs from the point of observation, enabled the Russians to acquire highly precise data on changes in the velocity of their target satellites.

The data from all sources was so exact that the Soviets quickly derived new formulas that enabled them to work out in advance the evolution of satellite orbits. They learned, for instance, that the latitude of perigee changed 0.35° during each 24-hour period and that atmospheric temperatures at great altitudes are much higher than had previously been supposed. As data was analyzed on the ground, the Soviets also learned within the first four months of the space age that a number of theories on X-ray and ultraviolet radiation from the sun were not valid. They discovered, additionally, that cosmic-ray intensity increased about 40 per cent within the satellites' altitude span. Like all good scientific experiments, the new Soviet investigations raised a great many more questions than they answered.

To obtain some of the scientific answers and, in effect, to fulfill Khrushchev's boast that "we can double, even more than double, the weight of the satellite," the Russians again astounded the West on May 15, 1958, when their beefed-up three-stage booster orbited a 2,925-pound "space laboratory." *Sputnik 3*, which I inspected in Russia, was an enormous, gleaming, cone-shaped spacecraft as meticulously engineered as the finest and most complicated surgical apparatus. I am no expert on welding, but it seemed to have been performed by a true craftsman. Not only was the interior crammed with an even greater variety of instruments than *Sputnik 2* had carried, but this newest and most superb of the world's satellites used revolutionary solar cells to provide the power for its electronic brains.

But long before the West properly evaluated *Sputnik 3*'s cargo of instruments—which eventually confirmed to the Soviets the presence of the earth's outer radiation belt—its deadweight of over a ton created the third and strongest jolt to United States pride and security. It was now obvious to nearly all that the combination of such a large booster, the Soviets' knowledge of nuclear weapons, and their ability to sustain life in orbit could presage a threat that was in some eyes far more sinister than mere scientific initiative.

While U.S. militarists pondered the awesome possibility of orbiting Russian nuclear weapons, the space age's sword of Damocles, educators and highest scientific councils speculated over the now rational and unpleasant possibility that an alien socialist culture might have overtaken what we had widely supposed to be an unassailable leadership. As the subject was discussed, from Washington conference rooms to regional teachers' meetings, one U.S. newspaper, the Minneapolis *Tribune*, took the unusual step of dispatching its top science writer, Victor Cohen, to Russia to try to find out what was going on in Soviet education and science. One calmer head than most belonged to America's distinguished Nobel Prize winner in chemistry, Dr. Harold Urey.

"Recent events," said Dr. Urey in 1958, "have emphasized to the people of this country in a most dramatic way that another country, whose political system we do not like and whose aggressive world policies cause fear among Western democratic countries, can and has produced scientific and engineering feats of a most remarkable kind . . . after some resistance on the part of the witch-hunting faction of the population we will settle down to understand our faults and shortcomings and do what needs to be done to correct the situation. But what needs to be corrected? Our security system? Nonsense! . . . Our support for science? Not in any immediate way. . . . The real problem that faces this country is the education and proper inspiration of our youth."

The first three Sputniks had shaken the American tree to its roots, and now two reactions set in. The first, in the U.S., was to begin the expensive process of redressing our grievous second-rate position on the ramparts of space. The second, in Russia, where the triumphant symbols of the Sputniks were even appearing on road signs, was to build on the strong momentum of a conspicuous and decided head start.

In 1957, there appeared in Russia a remarkable motion picture that the West was not to see for several years. Titled *Blazing a Trail to the Stars*, the science-fiction film is now considered by some to be the technological manifesto of Soviet space scientists; ever since, they have worked in the wake of its prophecy, at least in a general sense. It is obvious the film had the highest and most responsible technical advice. In it, with special effects extraordinary for Soviet cinemaphotography, bus-size segments of an orbiting doughnut are assembled in space by weightless construction crews.

The men work at the end of a tether and use sensibly designed torqueless tools. When assembly is complete, partial artificial gravity is mechanically induced by revolving the ring of the doughnut around a stationary central core. Finally, the charter inhabitants of the permanent space island—including several attractive women technicians and communications specialists, one with a pet kitten—are ferried and docked to the stationary center and transferred to the habitable ring. Viewers see the roomy, "shirtsleeve" interior, consisting of stateroom-like quarters, a miniaturized computer center, work areas, and a closed ecological-system space garden.

Later, a two-man moon rocket separates from the orbital platform and proceeds across cislunar space to effect a gentle, braking-rocket touchdown on the moon's desolate and alien surface. The climax of the film occurs as two spacesuited men gingerly, at first, then more confidently, step down and take the initial exploratory steps on the lunar surface. Finally, the two explorers, against a crescendo of background music—the Soviet version of "Pomp and Circumstance"—triumphantly embrace each other and do a stiff-legged dance of pure joy. What the climax lacks in literal fidelity, it makes up for in symbolism. Anyone who has wallowed for days in the incredible bulk of serious Soviet space literature cannot but be impresssed by volume after volume and treatise after treatise that reflect the magnetic attraction the moon and planets have for the Russians. Their long-recorded fascination with the celestial bodies of our ecosphere seems peculiarly compulsive, boldly romantic, and unabashedly obsessive. Both their national and scientific infatuation with the giant rocks of our solar system far transcends our own. While the extent of their attraction suggests strong psychological and even spiritual motivations, it is also firmly rooted in a knowledgeable and patient scholarship that often wearies the reader with its painstaking attention to technological minutiae.

That the moon and planets were the first and are the paramount Soviet scientific interest in space there can be no doubt. "All else is rehearsal," said Academician Parin. This philosophy is also intrinsic to a second film, *Before Man Steps into Space*, which the Soviets released in 1960, and in V. I. Levantovsky's book of the same year, *With a Rocket to the Moon*.

Thus, and again with dogged logic, the Soviets very early launched an attack upon the moon. Sergei Korolev once optimistically told an elite group of Soviet rocketmen that "the Soviet space program is designed so that, if possible, no shot will be repeated. Each will be a pioneer and build upon the work of the preceding one." The bold and prudently arranged chronology of the first three Sputniks was too perfect a progression to hold up in a new science. Yet the Russians attempted to sustain it by devoting nearly all of 1959 to initial probes at their target—the moon. They began the year on January 2 by sending *Mechta*, or *Luna 1*, into cislunar space. *Luna 1* had a massive third stage that weighed 3,245 pounds empty and was 17 feet long and 8 feet in diameter. Its spherical instrument canister separated after reaching escape velocity, but the two vehicles continued, virtually together, on a course that passed within 3,200 miles of the moon's surface after 34 hours of flight. Then they were drawn into the great magnetic whirlpool of the sun to circle our star as the first artificial satellite of the sun. *Luna 1* contained cosmic- and solar-ray counters, micrometeoroid detectors, tracking and telemetry transmitters, and one kilogram of sodium. The sodium was ejected when about a third of the way to the moon. It appeared briefly as a cloud of gas with the brightness of a sixth-magnitude star, long enough to establish an optical trajectory check and to provide the first data on the behavior of a gas in cislunar space. The flight substantiated the reliability of the big new Sputnik main stage, which now developed about 600,000 pounds thrust, inaugurated the era of rocket astronomy, and provided Korolev with a near-moon baseline from which to work in improving moon-directed guidance. The Soviet scientific secretary of the State Astronomy Institute, Victor Davydov, later said that *Luna 1*'s point of orbit farthest from the sun reached nearly to the orbit of Mars. At a speed of 19 miles a second, it circled the sun once every 450 days. Two

THE HEAD START 75

years later, Russia's artificial planet had traveled a billion miles. In January of 1961, it again reached the point in space where it had left the earth, but by then the earth had itself orbited to the other side of the sun.

Immediately after the Soviets announced the moon launch, the usual raft of stories and reports was circulated in the U.S. to the effect that the feat was a Soviet hoax perpetrated for propaganda purposes. At NASA's Jet Propulsion Laboratory in California, two emergency antennas, set for 70.2 and 183.6 megacycles, as announced by Moscow, were hastily rearranged. But the two antennas interfered with one another. Finally, two days after the launch, one was pulled down, and the 183.6-megacycle antenna confirmed the flight when it received—from an area just west of the moon—a two-tone modulation similar to that of Radio Moscow. A partial U.S. confirmation was also made by California astronomers of the Smithsonian Institution who photographed an object 450,000 miles out in space, which appeared as a small dot on a photographic plate. The Smithsonian assumed the dot was *Luna 1*.

Several congressmen, however, went on record that the Russian moon shot was a stunt or a hoax. A writer named Lloyd Malan got some quick publicity by publishing a book called *The Big Red Lie*. The hoax interpretation was never fully quashed, but most thoughtful people rejected that wishful thinking after Dr. W. H. Pickering, Dr. Wernher von Braun, and Dr. Homer J. Stuart testified before committees of the 86th Congress. Von Braun showed congressmen the evidence of the Jet Propulsion Laboratory's Goldstone tracking station together with a translation by Joseph Zygielbaum of a key *Pravda* article.

The Soviet success cost me $10. Back in August of 1958, after watching the explosion, after 77 seconds of flight, of our own first moon rocket—a combination of a Thor booster and Vanguard's second stage, known as *Pioneer 1*—I made a bet that we would get within 10,000 miles of the moon before the Russians did. Our second moon-bound rocket got only one-third of the way to the moon. Then, by a freak of luck, I managed to get past the security gate at Cape Canaveral for a close look at our third moon-bound rocket. I took the elevator up to the chemical-smelling eighth deck, signed in with a security guard, attached brass static arresters to the instep of each shoe, and gingerly climbed the steel stairway to the windswept ninth deck, where a family of misguided wasps was building its nest on the steel tower surrounding the U.S. lunar payload. For luck, I signed my name on the probe's white skin and gave Ed Bauer, a hard-hatted pad technician, a portion of a poem by Alfred Lord Tennyson. He later stenciled on the rocket the following words: "After it, follow it; follow the gleam." But the inscription must have jinxed it. While hopes were high, our third stage unaccountably failed to ignite. *Luna 1* got there first, and I lost my bet.

Due to Korolev and company, I also lost a bet that the U.S. would actually hit the moon first. Nine months after the *Luna 1* flight, Russian scientists aimed *Luna 2* directly at the moon, fired it, and waited in anticipation as the final stage strove for a highly precise burnout velocity of about 37,000 feet per second. They knew, of course, that as little as a one-second error in cut-off time could cause the 860-pound *Luna 2* to miss the moon. Unlike the orientation system on our Surveyor moon probes, which uses the bright star Canopus as a reference point, the Russians aligned their moon probes on the sun and the earthbound control station. On *Luna 2*, there was no facility for a midcourse correction. Accuracy depended on the launch phase of the flight. The Soviets, from their still-secret control center in the Crimea, flashed a predicted impact point to the Tyuratam cosmodrome just after the successful burnout. The message was received at Tyuratam with a burst of cheering. A second message, based on more accurate tracking data, was flashed after more than an earth day of flight, just five hours before impact. Upon receiving the corrected impact point, Air Force officers placed a red star on target on the huge photo of the moon that hangs in the cosmonaut mess hall at Tyuratam. *Luna 2* crashed where intended, destroying its instruments, near the crater Archimedes ($0°$ longitude, $30°$ N. latitude). The instrumentation it carried was basically similar to that of *Luna 1*. These two and confirming sources of data enabled the Russians to conclude that the moon has no magnetic field and no radiation belt.

The Russians have been fond of performing historic space missions on anniversaries of previous successes. On October 4, 1959—two years after Sputnik—they sent a canister-shaped, 614-pound photographic satellite into a highly elliptical earth orbit so calculated that its ascent to apogee would carry it partly around the moon. As *Luna 3*, with its cargo of a single, dual-lens 35-mm. camera, passed behind the moon at a distance averaging about 40,000 miles, the probe was initially stabilized by the ground based on lunar and solar references. Finally, and for a vital 40 minutes, its camera end locked on the gleaming orb of the moon. Its 8-inch and 20-inch lenses (at f/5.6 and f/9.5, respectively) photographed approximately 70 per cent of the back side of the moon. Readout procedures on the photographs were new and primitive, and the then-unknown problems of transmitting photographs across 240,000 miles of space resulted in a historic but slightly fuzzy first picture of the mysterious back side of the moon. A second readout was scheduled when *Luna 3* approached closer to earth, but it unaccountably failed to occur. Possibly, the thermal-control louvers protecting the transmitting apparatus were ineffective after some 50 hours in the bake-freeze of space.

The *Luna 3* photographs (the best one is a composite of nine) revealed that the back side of the moon was configured as differently from the front side as the Arctic is from the Antarctic. There are no large seas, or mares, such as the familiar and visible Mare Imbrium. Instead, there are a number of smaller ones set in a topography that is believed to be even more crater-pitted than the visible side. Near the center are several large and mysterious crevasses. The *Luna 3* photographs gave Soviet selenographers the rare privilege—virtually unknown to the modern world—of naming features on almost a hemisphere of an unexplored land mass.

In 1960, Russia started what we now know was a carefully programmed sequence of firings in preparation for manned flight. On May 15, *Sputnik 4,* a 10,008-pound satellite containing a dummy cosmonaut, was thrust into virtually a perfectly circular orbit 188 miles up. Inside the spacecraft was a 5,512-pound pressure cabin that was to be jettisoned in space and returned to earth for analysis. When Korolev's crew sent the electronic signal he had prescribed for the 64th revolution, the cabin separated properly; but when small jets flashed in the vacuum to stabilize the device, they worked improperly. The spinning cabin did not align for the desired re-entry attitude, and when its retro-rockets were fired they were pointed in the wrong direction. Instead of entering the earth's atmosphere, the cabin—or the eight pieces that were left of it—went into another orbit three miles higher, according to our NORAD tracking station in Colorado Springs. Two years later, one of the *Sputnik 4* remnants may have collided with a meteorite. A 20-pound Sputnik fragment was found embedded in a Manitowoc, Wisconsin, street, and in it were two rare minerals, wüstite and akaganeite, that had previously been found in meteorites reaching earth.

Soviet scientists did not consider the *Sputnik 4* failure to re-enter critical, and they did not attempt to repeat it specifically. Instead, they sent into orbit on August 19 two dogs, Belka and Strelka, other animal life, plant life, another dummy cosmonaut, and a two-camera television system trained on the biological specimens in a round, Vostok-like prototype of the first true Soviet "spaceship." *Sputnik 5* parachuted the dogs and the data to earth within seven miles of the intended target near the town of Orsk. The dog compartment separated at about 21,000 feet, and the Soviets were relieved to discover only minor ill effects in the canine passengers.

Sputnik 6, or *Spacecraft 3*, weighing five tons, contained a new Vostok automatic control system and an advanced new instrument bank for transmitting detailed physiological data on its two canine passengers, Pshchelka and Mushka. After launch on December 1, it worked perfectly in orbit. But after retro firing, the automatic stabilization system malfunctioned, causing a too steep and too rapid entry into the hard and abrasive molecules of the atmosphere. *Sputnik 6* and its cargo of living plants and animals perished in a fiery demise. Its thick metal shield glowed red, then white. Then it began to run molten. In an instant, it was turned into a fine ash that slowly drifted down to earth.

By now, Yuri Gagarin and Gherman Titov were entering their most intensive period of training for manned flight, along with ten other cosmonauts. For several weeks, news of the fiery failure was kept from them. Then one day, all cosmonauts were assembled where they could be watched by a battery of psychologists. There they were told the truth in every last detail, including the obvious fact that had one of them been aboard, he would have been incinerated. Someone broke the silence by jokingly suggesting that those responsible for *Spacecraft 3* could pay for it out of their own pocket. The tension was broken by general laughter. Titov reported later that two psychologists at the side of the room smiled and shook hands with each other.

Cosmonaut confidence received an even more desired boost on March 9, 1961, when *Sputnik 9* brought the dog Chernushka safely to earth after one orbit. Just sixteen days later, stepping up their schedule under pressure from the Project Mercury program in the United States, which had already scheduled Alan Shepard to become the first American spaceman, the Russians sent still another dog, Zvesdochka, for a successful one-orbit ride in the now-completely equipped Vostok spacecraft. The Russians had originally intended six orbits on their first manned try. This would have permitted recovery in a preferred area, but in 1960 Korolev revamped his plans and called for an initial manned flight of a single orbit. That is why the last two tune-up flights of *Spacecraft 6* and *7* were brought down after a single orbit.

When Zvesdochka got back safely, Korolev, in consultation with the highest level of the Soviet Academy of Sciences and with Khrushchev himself, gave instructions to prepare to launch *Vostok 1* whose passenger would be a short, modest young man from Klushino, Yuri Gagarin.

Prior to the Vostok series, the Russians also attempted to make their initial and long-cherished contact with the inner planets. After two failures with *Mars 1* and *2*, on March 12, 1961, they sent *Venus 1* within 60,000 miles of the brightest planet. *Venus 1*'s 1,419 pounds were packed with miniaturized cosmic-ray, magnetic-field, charged-particle, solar-radiation, and micrometeorite instruments, but Russia lost all contact with their ambitious probe when it was only 4,700 miles out. Unknown to the U.S. public, Russia quietly sent a delegation of scientists to the huge tracking station at England's Jodrell Bank Observatory. They wanted to see if they could detect signals from their planetary probe lost in space. Unfortunately for them, they were not able to. In exchange for this privilege, they later invited Jodrell Bank's director, Sir Bernard Lovell, to visit a Russian tracking station.

7
The Young Man
From Klushino

IN THE fall of 1959, just after Khrushchev's visit to the U.S. and his infamous shoe-tapping incident in the United Nations, a young MIG pilot flew south of the Arctic Circle from his airbase near the Arctic Ocean. Russian newspapers were filled with the pictures of the moon's far side that had just been sent back by the *Luna 3* moon probe.

"Once more," the pilot said later, "a tremendous wave of applause swept all continents . . . I realized I must wait no longer. The next day I made out an application asking to be placed on the list of candidates for space flight."

The signature on the application was that of Yuri Gagarin.

He was accepted, and the following spring got his first look at the Tyuratam cosmodrome, along with a busload of other cosmonauts selected for the rigorous training program. Riding with them was a Russian journalist named N. Melnikov, who took careful notes on the comments and reactions of the future cosmonauts.

The green bus picked them up near a railhead east of the Aral Sea. As they rumbled toward the cosmodrome for the 128-mile journey across the flat, sandy, and grass-covered desert, cosmonaut Gherman Titov tried to read to them from a book of Pushkin's poetry. Melnikov reports that only three of the group of cosmonauts appeared not to be paying attention.

Pavel Popovich and Valeri Bykovsky were trying to get a chess game going, and Yuri Gagarin appeared lost in his thoughts as the bus hummed along a new highway. As they were turning in through a security gate, Melnikov noted: "A light breeze is blowing over the cosmodrome. It can blow wherever it wants here, over the broad steppe. The sun is shining brightly, melting the light snow. The clear aromatic air quickens the blood, for spring is already blowing into the cosmodrome. . . . Quickly and noiselessly it sweeps away the snow, laying bare the damp earth. And then, in a single breath it dries the steppe and covers it with fancy carpets."

The first view Gagarin saw was a long rank of concrete buildings going up with the ubiquitous Russian construction cranes hovering and working above them. When they alighted from the bus, they met a young engineer who had been working there, he said, since 1953, fighting the summer heat and the winter cold, putting in telephone poles (which in Russia are affixed aboveground to vertical concrete posts), getting trucks unstuck from the clutching desert sand, and erecting the hundreds of rectangular buildings.

"Someday," he ventured "this city will be known as Kosmograd [spacetown] or Zvesdograd [city of stars]."

Gagarin and the others were shown their quarters in the "cosmonaut hostel," with its modern facilities, including a motion-picture projection room and a surrounding garden. The fledgling space pilots were then reloaded into their private bus and driven toward the launching area.

Gagarin was still silent as, Melnikov wrote: "the bus rushes along the smooth, broad highway . . . along the sides of the roads, settlements and the masts of high voltage lines flash past. The boundless steppe has already been settled and landscaped. The cosmodrome is a large outfit. Apparently, there's no end to it . . . it has the scope of space, if one might put it that way."

Gagarin, still silent, finally fixed his eyes on the horizon, where a projecting rocket was surrounded by the four steel girders of its service tower. Melnikov, somewhat quaintly carried away, observed that: "From a distance, it looked like a blast furnace. What a structure! It's gigantic! People are working up on top, on bottom, on the launching pad. No one is running around. No one is hollering or bustling about. Everything is done calmly, in a uniform rhythm. It's a measured, precise, business-like procedure. The cosmonauts look at all this with enraptured eyes and frequently exclaim 'Boy, oh boy!' They can't find any other words. The words will come to them later. The rocket stands at the ready and seems to be breathing like a living creature."

By this time, the spring of 1960, the Russians had already sent the dogs named Albina, Belyanka, Pestraya, Laika, Strelka, and Belka into space, and the waiting rocket the cosmonauts inspected was to carry yet another dog, a mongrel with a smoke-colored coat.

"What's her name?" one of the cosmonauts asked.

"Dymka [Smoky]," answered a doctor.

"What kind of name is that?" Gherman Titov kidded. "It's a heroic dog, and she turns out to have a name like Dymka! It doesn't suit her."

"Well," suggested the doctor, "think up a better name if you want to."

Different cosmonauts suggested such names as Svetlaya, Kosmicheskaya, and Laskouaya, but all were voted down. Then Yuri Gagarin pointed to the "star of hero" on the breast of a man nearby and said, "Let's call her Zvesdochka [Little Star]." Everybody liked the name, and Zvesdochka was the official designation of the dog atop the first rocket the cosmonaut trainees saw launched into space.

The occasion also marked the earliest indication of those peculiar qualities of leadership and respect from his colleagues that later and more decidedly came to be known as outstanding in Yuri Gagarin.

Late that summer, observers recorded another small bit of evidence. Once, when the cosmonauts were transported to the aircraft in which they were to train for weightless flight, their instructor for the morning session on the ground happened to be late in arriving. The cosmonauts loafed awhile. Then Gagarin, to avoid wasting time, suggested that he would start a discussion of the principle of the Kepler curve, which was the basis upon which an aircraft could simulate the weightless condition by flying a parabolic arc for 20 to 30 seconds. Gagarin, in effect, took over and became the instructor. This, and other evidence of enthusiasm and leadership, made a lasting impression on the other cosmonauts.

Later, the chief designer, whom we now know as Sergei Korolev, suggested to the cosmonauts that they rate one another. "In your opinion, who should be sent into flight first and why?" he asked.

The cosmonauts silently consulted their individual and group impressions and wrote down their recommendations. One of the young men named himself. "I should be sent," he wrote, then immediately added, "although I know that they will not send me. But my first name is 'Cosmic,' and this would sound good."

When all the recommendations were analyzed, the cosmonaut receiving the most recommendations—60 per cent, in fact—was Yuri Alekseyevich Gagarin from Klushino (a flax-growing Smolensk region of western Russia). His fellow cosmonauts gave as their reasons the fact that he performed his training duties well, "never loses heart," was a "fit comrade," a man of principles, "bold and steadfast," "modest and simple," and "decisive." Not long thereafter, Korolev agreed with the majority of the cosmonauts and named Yuri Gagarin as the "prime cosmonaut," to become the first man on earth to explore the lofty realm of space and the first man to orbit the earth.

The personable young pilot from Klushino was the son of a Smolensk peasant and carpenter, and the grandson, on his mother's side, of a metalsmith who could, as they said of him in Russia, "shoe a flea or hammer a flower out of a lump of iron." His mother, Anya, was a dairymaid on a collective farm.

Before he was yet in school, he recalled the response from his Uncle Pavel on the question of the possibility of life in space. "Who knows?" his uncle answered. "I think there must be life somewhere on the stars; it can't be that the earth is the only lucky planet among millions."

Yuri remembered the day, shortly after Germany invaded Russia, that the German infantry approached very close

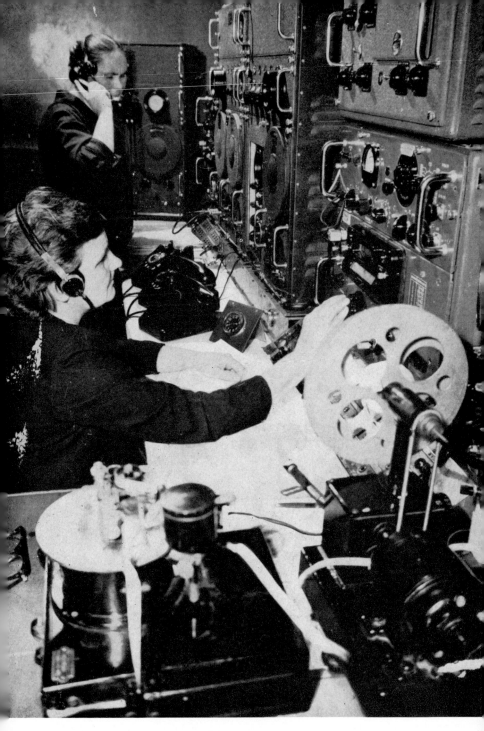

A scene in one of the U.S.S.R. Ministry of Communications' Control Stations near Moscow during the reception of signals from a Soviet satellite in 1962.

A rare photograph of the launching in 1961 of a large experimental rocket. When this photograph was first obtained by the West it caused considerable consternation. Because of the lack of anything to compare it to, the rocket appeared much larger than it actually was—which perhaps was precisely what the Russians intended.

while he was in school. That same day he saw his first airplane; a crippled Russian Air Force Yagg with red stars on its bullet-riddled wings landed in a nearby field. The pilot, Yuri recalled "unbuttoned his leather jacket and we boys saw the medals that gleamed on his tunic, and we understood immediately the price that had to be paid for military decorations. We boys all wanted to be brave and handsome pilots. We experienced strange feelings such as we had never known before." Finally, the Germans overran his village and occupied the Gagarin home while Yuri and his brothers and sisters built and lived in a dugout. During the next several months, thousands of German infantry, artillery, and tank troops plowed through the village on their way to the heart of Russia.

"We children," Yuri remembered, "did everything we could to annoy the Germans. We threw sharp nails and broken bottles on the road to puncture their tires, and we even stuffed rags into the exhaust pipes of vehicles that stopped for the night."

His entire village was without news of the front for months. Then a Red plane dropped leaflets telling of a great victory at Stalingrad (now Volgograd). When the Germans retreated, Yuri recalled with bitterness how they took two of his sisters with them and his parents followed the retreating columns until they were driven away.

After the war, Yuri continued his schooling at Gzhatsk, where he studied arithmetic, his favorite subject, with the counting aid of rifle cartridges instead of sticks. His early desire to become a pilot later received a boost when he studied physics under a Red Air Force veteran, Lev Bespalov, who first taught him how to use a compass. After six years of schooling, he took the train to Moscow to attend a foundrymen's school where, as a vocational-school boy, he wore a uniform with a peaked cap. He was interested in sports, especially basketball, and read translations of Longfellow's *Hiawatha* and novels by Victor Hugo and Charles Dickens.

He first read Tsiolkovsky in technical school and wrote a paper titled "K. E. Tsiolkovsky and His Theory of Rocket Motors for Interplanetary Travel."

D

In 1955, as a fourth-year technical-school student, he exercised his privilege of enrolling in a flying school that taught the theory of flight and related subjects at night. He took his first parachute jump five months later from an old PO-2 trainer. Girl as well as boy jumpers were in the plane with him, as he stood on the wing, clinging to the fuselage.

"Don't dither, Yuri," he recalled the instructor shouting artfully, "the girls are watching." Gagarin jumped and shortly thereafter made his first flight in a Yak-18.

"That first flight," he wrote, "filled me with pride and gave meaning to my whole life."

He received his ground-school diploma with a mark of distinction and began intensive flying that summer while living in a tent beside an airfield near Moscow. He even recalled some of the words of his instructors: "Flying is a business that never forgives the slightest mistake. . . . Your flying must be pretty to watch. . . . Strong nerves are more important than strong muscles . . . will power is not an inborn quality; it is something that can and must be acquired by training." By July, he had soloed in a plane known as the *Yellow 6* and, while waiting for subsequent flights, he recalled reading and discussing a book by the American test pilot Jimmy Collins. When he graduated, two years before *Sputnik 1*, he elected to enter military rather than civilian flying, and reported to his first station, at the huge flying field at Orenburg, a small city on the banks of the River Ural. Here, near a city that still used camels, he learned to fly jet planes. He was one of the first to take the Soviet Oath of Allegiance on January 8, 1956. Yuri Gagarin said he pictured himself on this occasion as a young man with Lenin's works in one hand and a rifle in the other. "The country," he said, "had entrusted us with arms and we had to be worthy of that trust."

During one of his leaves, he visited his girl friend, a nursing student named Valya, who was born in Orenburg a year after he was. He tried to impress her by cracking nuts with his teeth. "He has sharpened his teeth on the granite of science," Valya kidded about him to her mother. "He's been going to school all his life."

Later, when he transferred to winter quarters at Orenburg, Valya would sometimes smile at her platoon sergeant as he marched, but she had a hard time finding him; because he was short, he was never allowed to march in the front ranks. They decided eventually to be married. "Waiting is a great art," Valya wrote on the back of one of her photographs. "Reserve your feeling for the happiest moment, March 9, 1957."

When he began to fly MIG fighters, he discovered he was so short that he had difficulty orienting his plane properly for landing. He solved the problem with a cushion. Just before graduation, one of his companions, Yuri Dergunov, ran up to him with the news that *Sputnik 1* had been launched.

"That evening," he recalled, "as soon as we returned from the aerodrome, we all rushed to the club room and listened hungrily to reports on the movement of the world's first space vehicle and the speed that was so difficult to visualize, 8,000 metres a second."

"In another 15 years," a friend erroneously predicted, "and there will be a manned flight into space." None of them then knew how conservative this prophecy was, not even the very man who would prove it false.

Events immediately thereafter were all pleasant ones for Gagarin. He graduated as a lieutenant, and married Valya. His fellow officers toasted him with the traditional Russian "bitter": the wine is bitter and must be sweetened by the kiss of the bride. Just after the wedding, someone switched on the radio and they heard Premier Khrushchev announce that *Sputnik 2* was in orbit and that "two Soviet ambassadors, two stars of peace, are circling the world."

"And so, Nikita Khrushchev came to our wedding," laughed Valya.

Their happiness together was brief. Lieutenant Gagarin soon thereafter volunteered for Arctic duty, and Valya had to go back to Moscow to finish nursing school. While he was on Arctic flying duty, *Sputnik 3* was launched, and he read the first details in a late issue of *Pravda* that was finally flown above the Arctic Circle. He recalled making notes on the margin of the newspaper.

"I had an indefinite feeling then that rockets were going to replace aircraft," he recalled later. "I felt the time would come when our space pilots would bring back to earth samples of the moon's rocks."

Valya graduated from nursing school and joined him for the Arctic winter. The following spring, she gave birth to a 7½-pound girl, Lenochka, whom Gagarin carried home through the snow in an Army jeep.

Two additional Soviet space flights to the moon intensified his desire to become a space pilot. Once he heard a journalist on the radio ask Khrushchev a question: "When are you thinking of sending a man to the moon?"

"We shall send a man into space when the necessary technical conditions have been created," replied Khrushchev. "So far the conditions do not exist."

"These words," Gagarin remembered, "gave me no peace. I never doubted that I would apply for space flight. I was not afraid to begin my life all over again."

After he submitted his application, he was called up, along with several dozen other candidates, before a strict medical board under the direction of Dr. Yevgeny Alexeyevich. Many were turned away. "It's no use getting angry at medicine," the doctor told the rejects. "You can keep flying, only you can't go any higher than the atmosphere."

Gagarin passed a battery of tests, despite his constant worry that: "I was rather lacking in inches." At first, he told no one—not even Valya—that, as the final reviewing doctor put it, "The stratosphere is not the limit for you." He knew he would be transferred for special space training the day Valya baked a cake for his twenty-sixth birthday. But he was prohibited from telling her.

They flew together to the new cosmonaut training base where Valya and his daughter were to remain with him except for occasional trips home.

Gagarin's training already paralleled that of U.S. astronauts, with two conspicuous exceptions: The cosmonauts had intensive training in parachuting and spent hours, then days, at a time in a soundproof isolation chamber. Their ground school began with study of the space environment as explained by what he called "leading specialists" from Moscow who had worked with and analyzed Russia's numerous flights of animals into space. They also trained in heat chambers, the centrifuge, the vibrostand, and spacecraft simulators. Intensive physical training included gymnastics, volleyball, high diving and swimming, and working out with bars and dumbbells.

Gagarin particularly enjoyed parachute training. As a pilot, he had made his first jump several years earlier at Saratov and had made only four additional practice jumps before training for space flight. Now he practiced delayed jumps under Russia's Master of Sports, Nikolai Konstantinovich, who had once made a record delayed jump of 5,270 feet.

The first time he met the chief designer, Sergei Korolev, Gagarin described him as a "broad-shouldered, merry, witty man, a real Russian."

"It is hard going," Korolev told the cosmonaut class, "but you must go through it to be able to stand it up there." Then the designer took them to see the round spacecraft he had designed. It looked, at first glance, like a huge cannonball covered with asbestos. Portholes the size of a catcher's mitt were covered with fireproof glass. Korolev explained that inside cabin temperatures would vary from 15° to 22° centigrade. "You see," Korolev said, "the outer skin of the ship and the pilot's cabin are coated with reliable insulation that will keep the ship from burning up during re-entry." Korolev also explained the special metals, glass, plastics, durable fabrics, and fireproof varnishes the 10,417-pound spacecraft contained. The chief designer also told them something they did not then know: the first manned flight was scheduled for one orbit around the earth.

Gagarin and the other cosmonauts climbed into the cabin, which, he said, was larger than an aircraft cockpit. He got his first look at the bank of pilot's instruments to his left, which contained radio switches and dials and environmental-control regulators, and, on his right, the "orientation-control handle" (stick), food locker, and radio receiver. Directly in front were other needle instruments, an electric clock, and a globe of the earth designed to revolve synchronously with the orbital path. The complicated contoured seat especially interested him because it contained such safety devices as a restraining harness, a catapulting device, parachutes, and emergency food and water. The sphere also contained a television camera. The spacecraft, he was told, was so designed that one landing mode called for the entire spacecraft, containing its pilot, to parachute to land or, if necessary in an emergency, to water; the secondary means of landing was to catapult the seat so the pilot would descend alone on his own chute.

"Everything was light, durable and portable," said Gagarin. "I left the cabin in silence and in silence made way for my comrades. . . . I could not find words to express my admiration."

As training progressed, Valya tried various ways of finding out where his demanding and secret training would eventually lead him. Once, he admitted, she tricked him into saying: "I'm going on a trip into outer space. Put a clean shirt in my bag." Although she treated the remark as a joke, she must have sensed the truth, for she asked no more questions. Valya also knew, of course, that on December 1, 1960, a practice spacecraft containing two Soviet dogs and numerous other animals had been lost due to a faulty retro attitude and too steep descent path. The entire spacecraft had burned up on re-entry. But the loss of the spacecraft was a cause of some initial concern only. "The bitter aftertaste produced by the loss of Pshchelka and Mushka, which we were afraid to admit to ourselves but which existed nevertheless, completely vanished as soon as we learned of the next successful orbit and recovery of the spacecraft."

Gagarin sensed the time for manned flight was growing near on August 19, 1960, when Russia successfully lofted two dogs, Belka and Strelka, into eighteen orbits inside Korolev's brainchild. This event, Gagarin said, "demonstrated to all of us the reliability of the ship we were studying." But he still did not know for sure who the first human pilot would be, although he was now being given some preferential treatment and training as "the senior of the group."

After the fourth satellite spaceship returned to earth with the dog Chernushka still alive, Gagarin heard Khrushchev say on Radio Moscow he was very confident that soon the first spaceship with a man aboard would be launched. Gagarin said: "I felt at once elated and a bit uncanny."

The young man from Klushino was familiar with U.S. preparations to launch a man under Project Mercury. There were piles of American magazines and newspapers in the cosmonaut lounge, and they were often looked at. He had seen in *Life* a picture of the orbiting chimpanzee Enos, and he had read translated UPI reports on the selection of the "prime" U.S. astronauts, Alan Shepard, Virgil Grissom, and John Glenn. He did not, however, consider the planned, first-manned ballistic shot of Project Mercury a true space flight. He reacted to the proposed flight somewhat as many Americans reacted to Soviet "space spectaculars." "Apparently," he said, "it was aimed at creating a sensation." For the American astronauts proposed for space flight, he had professional admiration, calling them the "courageous American fellows."

Early in 1961, the Gagarins' second daughter was born, and the parents gave her a spring name, Galochka. "The breath of spring could be felt in the air," Gagarin wrote, as he went out to the launch pad to watch the last preman launch of Korolev's cannonball.

He definitely knew now that he had been selected for the first flight. He had imagined during the launch of the dog he had named Zvesdochka what it would be like to take the dog's place, and his friends were already congratulating him. Valya, however, had still not been told definitely by either her husband or Korolev. Then one night, after they put the children to bed, he told her that he might be the first man to go into space. "But why you?" she asked as, Gagarin recalls, her face grew serious and her lips began to tremble. They talked throughout the night without closing their eyes. In the morning, Valya told him: "If you are sure of yourself, go. Everything will be all right."

Gagarin's friends had often repeated a Russian saying that

"the East is closer to the sun than the West," and he learned in early April that his spacecraft had been named the Russian word for East, *Vostok*. The night before he left for the cosmodrome, he was toasted at a party. One man said: "When you come back from outer space don't be conceited. Don't snub us. Continue to be as modest as you are today."

Gagarin replied that he was prepared to make the flight "with a clear conscience and a great enthusiasm."

He visited Moscow one more time. Then Gagarin and several backup cosmonauts, including his number one "double," Gherman Titov, plus suit technicians and a physician, were flown to Tyuratam. As the plane flew across the great central Russian plain, he said, he was comforted by words he recalled from his early days as a foundryman apprentice. His supervisor had said: "Fire is strong. Water is stronger than fire. Earth is stronger than water. But man is stronger than all of them."

The plane came within view of the Kazakhstan steppes, which Gagarin described as "wide as the ocean" near Tyuratam. On April 4, the plane touched down at the cosmodrome airstrip for Gagarin's appointment with destiny.

8
The Gagarin Flight

THE unofficial race between the U.S. and the U.S.S.R. to launch the first man into space was close. In early April of 1961, a few days before Gagarin's flight, I was living in Cocoa Beach's Holiday Inn, getting material for *Time* magazine's forthcoming cover story on the approaching flight of the first American in space, Alan Shepard. Although we knew that Russia had been testing its Vostok spacecraft and had propelled a seven-ton payload toward Venus—indicating a truly massive booster thrust—none of us, of course, could even guess at the scheduled launch of the first Russian manned flight. Uppermost in the minds of the Mercury-Redstone launch crew and of the few astronauts I talked with was the definite possibility that we might be first in space. "Sure, we'd like to be first," Gordon Cooper told me over dinner. "Everybody here at the Cape is working his tails off."

Evidence of this was apparent not only in conversations with rocketmen but also in the visible signs of an imminent launch. On Cape Canaveral's Pad 5, the old reliable Redstone rocket, MR–3, was already in position. At its tip was the Mercury spacecraft with the name Alan had selected, *Freedom 7*, already painted on its side. The latticework linking *Freedom 7* to its escape rockets projected beyond its nose had just received a fresh coat of bright scarlet paint.

Even the scene off the Cape was one of near-feverish activity. Network and pool press representatives had already checked into the Holiday Inn, where they were quickly converting motel bedrooms into makeshift offices with

Teletype machines, bulletin boards, and direct-line telephones. The outside of the motel was festooned with the serpentine coils of black cables that dangled like spaghetti from the second-story landing. Even traffic on Cocoa Beach's new four-lane thoroughfare seemed to move at a livelier pace. A song group known as the "We Three Trio" was already rehearsing a special song called "Space U.S.A."

Alan Shepard himself had already checked into his robin's-egg-blue quarters inside Cape Canaveral, along with his backup pilot, John Glenn.

Simultaneously, ten time zones around the world, Yuri Gagarin had now been separated from all the other cosmonauts except his substitute, Gherman Titov, and they were living the final hours of preparation in a small brown cottage that also had walls of robin's-egg blue. The day before the flight Gagarin listened to "soothing music" on a tape recorder. That evening he played a short game of pool; then he and Titov dined with their doctor, Yevgeny Anatolyevich.

Just before bedtime, Sergei Korolev dropped in. "In another five years," said Korolev, "we shall be able to fly into space as we now fly to resorts."

After Korolev left, Yevgeny Anatolyevich suggested sleep, and motioned to the pocket of his white gown. "Perhaps I should help you with your sleep?" he hinted, half in fun. Gagarin declined the soporific and climbed into his three-quarter bed with an old-fashioned iron bedstead. When Anatolyevich tiptoed in thirty minutes later, he found Gagarin lying on his back, fast asleep.

The Vostok designer and launch director, Sergei Korolev, however, could not sleep. For a while he read the latest issue of *Moskva* magazine; then he, too, tiptoed into Gagarin's bedroom, at 3 A.M. In the distance, he could hear the rumble of trucks heading for the launch pad. Gagarin was fast asleep; the designer tiptoed out again.

Yevgeny Anatolyevich awakened the pilot from Klushino at 5:30 A.M.

"How did you sleep?" asked the doctor.

"As you taught me," said Gagarin.

Gagarin did setting-up exercises, then had breakfast from

tubes of chopped meat, blackberry jam, and coffee. Assistants began the long process of collecting last-minute medical data and dressing him for flight. First, he put on a suit of sky-blue thermal underwear, then his orange-red space suit. He received last-minute instructions from the parachute instructor, Konstantinovich, who checked him out again on how to catapult himself, if necessary, free of the spacecraft just before touchdown. Finally, Sergei Korolev entered. Gagarin observed: "It was the first time I ever saw him careworn and tired—apparently the result of a sleepless night. And still a soft smile played on his firmly compressed lips. I wanted to embrace him as I would my father."

After receiving the chief designer's advice, Gagarin put on his white inner helmet and hard white outer helmet and was escorted to a waiting blue and white bus. Soon the bus sped along the highway.

In a matter of minutes, he was in sight of the launch pad located on the desert steppe. Several days earlier, a heavy-gauge-steel railroad car known as a transporter-erector had backed its way to the pad. Its cargo, jutting far out over each end of the flatcar, was slowly hoisted by the erector to a vertical position where it was securely gripped by four steel towers. The white, segmented rocket had twenty main engines, each connected to a flared nozzle about 30 inches in diameter. In addition, there were twelve smaller vernier engines. The main cylindrical booster supported at its tip the *Vostok* spacecraft, whose bullet-shaped aerodynamic shielding was 33 feet long. The payload was separate from the booster and was joined to the upper fuel-tank section by metal struts.

Beginning about halfway down the booster were four conical-shaped engine compartments, each of which flared to its array of four booster and two vernier engines. Each compartment contained a single triangular white fin.

When Gagarin stepped down from the bus, the weather reassured him; only a few distant feathery clouds were visible.

Greeting him at the launching site were the chief designer and the chief theoretician, standing close together as always.

The other cosmonauts as well as assembled dignitaries and scientists were gathered nearby. Gagarin walked over to the Chairman of the State Commission for the First Space Flight and said formally: "Senior Lieutenant Gagarin is ready for the first flight in the spaceship *Vostok*."

He looked up at the ship, which he felt was "more beautiful than a locomotive, steamship, plane, palaces and bridges all put together." Then, in the Russian fashion, he made a formal speech that would have sounded stilted and grandiloquent to Western ears. Most of Gagarin's comments, in fact, suggest a bashful and somewhat naïve man of the country.

"Dear friends, intimate and unknown, fellow citizens, people of all countries and all continents," he began. "All I have done or lived for has been done and lived for this moment. . . . To be the first in outer space, to meet nature face to face in this unusual single-handed encounter—could I possibly have dreamed of more."

As his speech lengthened, he saw Korolev give a furtive glance at his watch. "I am saying good-bye to you, dear friends," Gagarin continued, "as people always say to each other when leaving for a long journey. I should love to embrace all of you, friends and strangers, near and far . . . see you soon!"

Ninety minutes before liftoff, Gagarin rode the inclined elevator fifteen stories to near the top of the steel-girded rocket. He climbed inside *Vostok*, and assisted in the closing of his hatch. All was silent. The cabin, he noticed, still smelled of the spring fields.

"Hello, Earth, I am a cosmonaut" he said, testing his communications. He checked his cabin temperature (19° C) and humidity (65%) and pressure (1.2 atmospheres) and listened to the toll of the countdown.

At the command "Liftoff," Gagarin said, "Off we go. Everything is normal," and glanced at his watch. It was 9:07 A.M. Moscow time, 4½ hours since he had been awakened. He heard the roar below him, about as loud as jet engines during an airplane takeoff.

At this point, Gagarin could not see outside since the

long, pointed shell of the nose cover still enveloped the spacecraft. But after the contrail spun out and the rocket had passed denser strata of the atmosphere, the nose cover was automatically jettisoned.

"How magnificent," he said of the view, despite his efforts to remain calm. "I see the earth, forests, clouds. . . ."

He felt some vibration and G forces but was relieved that the G forces were less than he had encountered in the centrifuge training. By the time the orbital thrust rockets fired, the cabin temperature had risen one degree and cabin pressure had dropped to exactly one atmosphere. When he first felt himself weightless, he found the condition uncomfortable, "but one quickly adapts . . . I felt an extraordinary lightness in every limb." He rose from his seat and watched a stray drop of liquid that had escaped from a feeding tube float about, then finally come to rest, "like dew on a flower," on a glass porthole. As if to heighten his reveries, the ground station began to play his favorite song, "The Amur Waves."

Ground instruments now showed *Vostok 1* to be securely in orbit on an inclination of 64° 57′. This course took him over southern Russian mountains and over India. Forty-two minutes after launch, the automatic orientation system began to function by aligning *Vostok* with the sun. It remained in this alignment as he passed, at nearly 18,000 miles per hour, over Cape Horn. He practiced eating from tubes and listened to the sound of his own heart, which he was relieved to find had now settled to near normal in the state of weightlessness. But because of the leads and instruments hooked to his body, those on earth knew more about his physical condition than he did.

So far the automatic stabilization system was working perfectly, and it was not necessary for him to take over control manually. One hour and eight minutes into the flight, while he was orbiting near Mt. Kilimanjaro in Africa, the controls automatically realigned the ship for the firing of the retro-rockets. Thinking of the superheating of his spacecraft during the approaching re-entry, Gagarin felt some apprehension. Retros fired 1 hour and 18 minutes after liftoff, precisely on time. Gagarin reported: "The descending ship began to enter the dense layers of the atmosphere. Its external skin rapidly became red-hot and through the porthole filters, I saw the frightening crimson reflection of the flames raging all around the ship. But in the cabin it was still 20° centigrade despite the fact that the ship was hurtling toward earth like a ball of fire."

Keener apprehension came when he felt powerful G forces pressing him against his contoured seat as, simultaneously, the round spaceship began to rotate. Fortunately, the potentially dangerous rotation, caused by the first and uneven harsh friction of the atmosphere, quickly damped itself out.

When he saw the silver streak of the Volga River flash below him, he recognized the airfield where he had once taken flying lessons. Soon, as the ship descended under its red-and-white parachutes, he recognized the thatched houses and livestock barns of the nearby village of Saratov. The precise target had been an airfield southwest of the town of Engels. The fact that he had missed this only slightly testified to the remarkable accuracy of the Russian automatic equipment. The heavy round ball hit in a plowed field, rolled over, and stopped, as the main chutes fell off to one side, not too far from a deep ravine.

The first thing Gagarin saw as he stepped out was a woman and a girl standing beside a spotted cow. Still wearing his flaming orange-red space suit, he started walking toward them. The woman and the girl approached him also, but as they grew closer, their steps understandably slowed at the strange spectacle of a seeming humanoid in a devil's suit.

The woman, Anya Takhtarova, the wife of a local forester, asked uncertainly, "Did you really come from outer space?"

"Just imagine it. I certainly have," Gagarin replied.

They were soon joined by a group of machine operators, who shouted his name, which they had learned from reports on Radio Moscow. Then a group of soldiers arrived to greet him for the first time with his new rank of major. He had been promoted in flight, skipping the rank of captain.

After the soldiers helped him out of his space suit, he inspected *Vostok 1* in his sky-blue underwear. "The ship and equipment," he said later, "were in good enough condition to be used for another space flight. Tremendous joy surged through me."

As the triumphant national hero was wheeled away from the plowed field near Engels, word of his safe recovery was flashed throughout the world. Soviet officials, now fully cognizant of the worldwide impact of the first Sputniks, went all out in providing most flight details, except the specific nature of the booster and spacecraft. Their expansive treatment of the flight, as evidenced in Tass, *Pravda*, and *Izvestia*, had begun as soon as he was in orbit, reflecting the confidence of Korolev and company.

"The cosmonaut, Major Yuri Gagarin," Tass first reported with restraint, "withstood the period of the spaceship's injection into orbit remarkably and at present he feels fine."

Subsequent announcements confirmed his sequential presence over South America, Africa, the U.S.S.R., and finally Tass confirmed in nontechnical language that "the braking motor installation of the spaceship was activated in accordance with the flight program, and the spaceship *Vostok* with Gagarin aboard began its descent from orbit for a landing at a predetermined point in the Soviet Union."

But once the young major was safely recovered, officials and Soviet news media pulled out all the stops in an effulgent, occasionally mawkish, always star-struck account of his widely enthusiastic reception by the Russian people. He had become the first of the space age's instant heroes.

As if on cue, Soviet politicians were quick to convert the universal acclaim into their newest instrument of expressing the avowed national purpose. The Soviet Government and party hailed "the new era in the development of mankind" and issued an appeal for all countries "to effect universal and complete disarmament." Khrushchev trumpeted predictably, "Let the capitalistic countries try to catch up. . . ." For the rotund former peasant, the event marked a high point of personal prestige. "You have made yourself immortal," he told Gagarin on the phone, but he may have been thinking the same thing of himself, recalling his bitter battle to give top priority to space. "I will meet you in Moscow," the Premier told Gagarin. "You and I and all the people will solemnly celebrate this great feat. . . . Let the whole world see what our country is capable of. . . ."

Solemnity was hardly the word to describe the mood of Gagarin's reception. The Soviets at first flew him off by helicopter and secreted him in a prearranged cabin on the banks of the Volga for his physical examination, debriefing, and rest. After a walk along the riverbank, a game of pool, and a good night's sleep, he arose refreshed to begin his first postflight day with his accustomed setting-up exercises. If he had needed conditioning for space flight, he was to find out —as Alan Shepard was soon to discover—that he needed even more conditioning for the physical and emotional strain of being a besieged symbol of suddenly rampant and powerfully emotive national feelings. He saw Korolev first and was pleased to note that he now seemed relaxed as, in the Russian custom, they embraced and kissed. Next, Gagarin gave a long, part-personal, part-engineering account of his flight to assembled scientists, then repeated portions for the benefit of assembled newsmen.

"The newspapers made me happy," he later reported, "and at the same time embarrassed me. To become the center of attention not only of one's country but also of the whole world is rather burdensome."

Both the happiness and the burden grew greater as he boarded an Ilyushin-18 plane for "Mother Moscow." Approaching Moscow, the IL-18 was escorted by seven sleek MIG jets similar to the ones he had flown in the Arctic. Over the capital, he said: "I looked down and gasped. The streets were literally packed with people." Upon landing at Vnukovo airport, he saw other thousands of shouting people awaiting him. A red carpet, fifty yards long, stretched from his plane door to the flower-bedecked speakers' dais, where the most powerful men in Russia, in and out of uniform, were assembled. "I had to go, and go alone. And I went. I never felt so nervous in all my life." He felt one of his shoelaces come untied and had a horrible vision of himself tripping and sprawling on the red carpet. But unaided, he reached the platform, where Khrushchev, Valya, and his parents, among others, congratulated him. Valya was crying as she clutched a bouquet of roses, a gift from Nina Khrushchev.

After the speeches, there was the parade and the august formality of a densely packed Red Square. His ovation was strong and sustained. He remembers blushing when Khrushchev announced he was awarding him the title of "Hero of the Soviet Union." Gagarin unabashedly pictured himself with a tunic full of medals, the Order of Lenin, the Gold Star, and others.

After the reception in the Grand Kremlin Palace, he faced the press conference of the Soviet Academy of Sciences. Then there were even more speeches and decorations.

MOSCOW'S DANCING ON AIR headlined the Miami *News*. GAGARIN PAID HOMAGE DUE ONLY TO CZAR, blazoned *The Chicago Tribune*.

The tumultuous wave of celebration engulfed Gagarin again when he visited the satellite countries, Czechoslovakia and Bulgaria and, later, many of the principal cities of the world. It was a wave that was a long time in subsiding.

Gagarin decided while still in orbit that he would visit Konstantin Tsiolkovsky's grave in Kaluga. This he did, and laid the cornerstone of the new Tsiolkovsky museum that was still under construction when I visited Kaluga in the summer of 1966. He also visited his home town of Klushino, where he saw the fading evidence of the German occupation, during which he had seen his first Russian airplane and pilot. And he saw the rye and flax and heard the woodcocks and nightingales known to him from his youth. The young man from Klushino had finally come home.

9
The Poet of Space

In the heady euphoria following Gagarin's conspicuous premiere in *Vostok 1*, Russia was in no hurry to proceed with other space shots. In fact, not for four months did a rocket of any kind roar up from either Kapustin Yar or Tyuratam. It was a period of quiescence in which the Soviets knew the United States would take little comfort. Russian leaders had already learned that although few had paid attention to their preflight explanations, this same detachment had contributed to the belated but overwhelming attention generated by each record-breaking accomplishment, and they had abundant evidence that the flight of *Vostok 1* had become a spectacular cynosure of the entire world. The delayed-reaction pattern was one they grew to like; they began to lean more and more toward a shoot-now, talk-later philosophy.

Politicians, scientists, and the Soviet people drew out and relished the national emotional reaction to man's first penetration of space. They made the shy and uncomplicated Gagarin not only an instant knight but something of a saint as well. This highly emotional reaction with its religious overtones has been underestimated in the West. In fact, it has been basically incomprehensible to a society whose concept of man's destiny is firmly established and which thus far has felt no compelling need to create new ones or to enlarge man's traditional biblical destiny to include a destiny in the cosmos.

The new, evocative self-identity of the Soviets and their exhilaration before the starry throne of space suffered no restraint whatsoever when, on May 7, 1961, Russian newspapers carried accounts of the successful ballistic flight of Alan Shepard, who had ridden inside the Mercury spacecraft for fifteen minutes. The response to the Shepard flight reflected the magnanimity of the victor. Previously, Gagarin and Khrushchev, together and separately, had told their countrymen that a passenger on a ballistic trajectory could hardly be called a spaceman, nor his flight designated a true space flight. The seed was self-watering and was now nourished by the event itself. When Shepard was successfully recovered, the Russians came to the public forum to praise him. Khrushchev cabled President Kennedy: "The recent outstanding achievements in man's mastery of outer space open unlimited opportunities for the cognition of nature in the name of progress. Please convey my congratulations to flyer Shepard."

Gagarin read of the flight while visiting his home and was equally generous. After noting that Shepard's flight lasted only one-sixth as long and covered only one-ninetieth the distance of his flight, he commented: "I vividly imagined how the American must have felt in the cabin during the delays in the countdown. . . . Alan Shepard did everything American science and technology enabled him to do. He is a brave man. I should like to shake hands with him and to wish him and his family every success."

Unknown, unquoted, and virtually unspoken of on the subject of space was the man destined to become the Soviet Union's second space pilot, a heavy-browed, curly-headed, poetry-loving Russian named Gherman Stepanovich Titov. "Gherman" is a highly unusual name for a Russian, something like "Percival" or "Roland" for an American. Titov came by his odd cognomen because his father, a village schoolteacher in the Altai territory of southeastern Siberia, named both of his children, Gherman and his sister Zemfira, after characters of his favorite author and poet, Alexander Pushkin.

Russia's second spaceman was quite different from Yuri Gagarin. Both were products of the Russian village; both had early fallen in love with the uniform of the Red Air Force and had developed a consuming ambition to become a pilot.

But their characters were in marked contrast. Gagarin, in his simple idealism, his humility, and his open-faced personality which struck few antagonisms, was akin to the wholesome Boy Scout image that Americans were later to recognize in the character of John Glenn. The astronaut most like the more sophisticated, well-read, and articulate Titov is Wally Schirra. Both are outspoken, have to watch their temper, and, as pilots, are superbly coordinated and coolly professional. Those who know Titov describe him as occasionally "hot-tempered," "stubborn," and "overly intense." Yet his competence in the cockpit was so obviously superior that it far outweighed personality traits in most people's assessment of him.

Titov had a slight eccentricity of health that escaped early examination and that, Russian doctors now admit, would have eliminated him from space-pilot training had his examinations been more thorough and strict, or at least more oriented to the new science of bioastronautics.

As a child, he was once caught in a blizzard and had fainted on the roadside, and a broken wrist sustained from a bicycle accident caused him years of apprehension, which he tried to overcome by hard-driving arm exercises. After his youthful accident, he kept the sharp pain in his wrist from his family and mistakenly scorned a splint or bandages. He tried, instead, to overcome the weakness he felt in his wrist by secret exercises. He once exercised his injured wrist so hard that he vomited from the pain. Later, after his selection as an aviation cadet in early 1953, he was so afraid his wrist would wash him out as a pilot that he got up an hour early each morning to work out alone on the parallel bars.

It was during the period he was taking his lonely workouts in order to remain a pilot that his unpredictable temper almost cost him his career. Titov took a dislike to a certain colonel, who also had a quick temper. One day the colonel made a remark with which Titov disagreed, and the brash young cadet shouted back at the officer, "adding certain remarks," Titov commented later to two British journalists, Wilfred Burchett and Anthony Purdy, "which were incredibly foolish for any cadet. . . . He fairly exploded on the spot.

The difference in our tempers was that he was a colonel and I was a cadet. He shot to his feet and demanded the strictest of punishment—which would have meant an immediate end to my flying career. I had committed an unspeakable sin of military protocol, and this man demanded—and by virtue of his rank, could get—maximum punishment."

Titov has his flight instructor to thank for saving his career. A bold Captain Korotokov addressed the corps of officers, including the still-fuming colonel, in what must have been both an eloquent and a desperate speech. "I dare say, sir," the captain said to the colonel, "that I know young Titov as well as anyone in this room. We have to watch for his tendency to extremes. Perhaps he has not yet had sufficient time to realize that as a future officer who is to give orders, he must learn to accept orders—pleasant or not—and to carry them out instantly. His fault at the moment is in slighting orders but this is not deliberate. Why should we punish him because he defends his point of view? Is this not what we want in our future pilots? There are so many positive things in Titov's character, his sober independence of thought, his will power, his discipline in every other respect, his superb flying ability, his creativeness. These things are too good to be thrown away."

Captain Korotokov's calm and remarkable plea on Titov's behalf persuaded the colonel and the corps of officers, and Titov was allowed to remain a cadet. As soon as he could, the captain got Titov to one side. "Titov," he told him, "you are an ass. You were wrong in there today and you're lucky to have gotten away with that stupid temper of yours. Have faith in yourself! But have faith in others, too!"

Thereafter, as Titov gained flying experience, he resented but eventually mastered the constant attention to the mechanical and engineering side of aviation. His basic romantic attitude toward flying and his penchant for the poetic popped out whenever he had a Yak trainer to himself, a bright sky above, and sunny clouds to circle. He fell in love with flying under the eyes of a girl who, for a long time, watched and admired him from a distance.

Despite his growing proficiency and daring in the air,

Titov was inordinately shy of girls—so shy, in fact, that he made only a tardy and perfunctory appearance at scheduled officers' dances. Often he read Pushkin alone in his quarters until the dance was nearly over. But on one such late appearance, a girl who had the beautiful name of Tamara stopped him in his tracks with a dazzling and deliberate smile. Tamara turned out to be the unit cook; often, from the privacy of the kitchen, she had watched him eat food she had lovingly prepared for his particular enjoyment. Titov fell in love immediately. He recalls: "I found in Tamara the wonder of life in a vein I did not know existed. . . . She stripped away from my life the shell that stood between myself and the deep river of emotion we would soon share."

He proposed to her two months after they met. At the wedding at the airbase, Titov's regimental commander surprised him by giving him the key to a private apartment located on the field. Their idyllic life was shadowed only by Tamara's constant and growing fear of her husband's flying.

Titov, thanks to Captain Korotokov's faith in him, eventually stood at the top of his class academically and became the first pilot to solo in a jet. Titov called the MIGs "beautiful machines, gleaming in the sun, their wings cut sharply back; every now and then they screamed over the training area like some product of aerial sorcery." When Titov finally graduated, his instructor prophetically wrote of him: "Particular attention should be paid to this young man. He will develop into a first class pilot. He flies boldly, confidently."

One cold October night in 1957, a night Titov described as "a flyer's night," he stood alone on the airstrip at Leningrad. He gazed absently at the stars, watched a cruising jet, then suddenly was startled to see a strange "crisp and clean" light moving against the black vault of the sky. The experience of seeing something awesome and strange that he could not identify brought him vague displeasure and he retired to an uneasy sleep. The next morning he learned from the radio and the excited yells of his fellow pilots that the mysterious object he had seen by chance was *Sputnik 1* tracing the moving line of history through the immutable constellations.

"Never could I have imagined," he said later, "that my

own reach above the earth would in only several years' time bring me to the heights of that first satellite."

A few months later, Lieutenant Titov was summoned to his commander's office. There, two strange colonels threw questions at him in rapid order. How did he like jets? How did he feel about speeds many times the speed of sound? Had he kept up with progress in rocketry?

He was extremely puzzled and curious about their purpose, but he confined himself to giving accurate, straightforward answers. "You do not ask brash questions of colonels," he said. "I had learned that lesson well."

Then the truth dawned on him like a bursting star—they were grilling him so pointedly because he was under consideration for space-flight training.

"There is a new field of aviation about to open up, Lieutenant," one of the colonels finally said, slowly and deliberately. "It will be extraordinarily difficult and dangerous. The training will be backbreaking. It will go on day and night until it drives you crazy. . . . There is no time to think it over, Titov. I ask you; are you interested?"

Titov stood up. "Yes sir, I am," he replied instantly.

From that day forward—the same day he learned from Tamara that he was to become a father—he thought of little else but a now-magic word one of the colonels had casually used, "spaceships."

Titov was sent to an Air Force hospital in western Russia, where he met other cosmonaut candidates and submitted to a barrage of tests in pressure suits, pressure chambers, on the centrifuge, and in a jerry-rigged bus seat ingeniously designed to "rattle a man's eyes in their sockets." He was baked under a blaze of heat lamps, asked to sit with his feet in ice water, had his body hooked up to live electrical wires, suffered explosive decompression and oxygen deprivation, run on a treadmill, and probed with hundreds of questions by a battery of psychologists. He lost his temper only once, when he flared up in front of a psychologist at the senselessness of some of the "silly questions."

But thanks to another fortunate intervention, this time by a prominent Soviet doctor, Eugene Alexeovitch, he was not eliminated. His stubborn streak carried him through even harsher and more merciless competition. By forcing himself to endure extreme physical punishment that he could barely stand, he won the grim satisfaction of seeing a few incredulous doctors raise their eyebrows. "All we want," one man told him, "is a quart in a pint pot."

When his selection for detailed space training became official and he was allowed to tell Tamara, his vivid imagination constructed a sort of cloud-nine world of the spaceman—a view that was abruptly shattered when he and his pregnant wife reported to the cosmonaut training field in White Russia, a place that turned out to be a routine airbase with drab barracks, obstacle courses, and drill fields. The early lack of sophisticated, far-out space apparatus distressed him. "We could have been a team training for the Olympics," he at first observed.

The cosmonaut training and spaceship orientation program as seen through Titov's eyes provided a more complete picture than that given by the relatively complaisant and star-struck Yuri Gagarin, whom he now knew and with whom he often played ice hockey and basketball. Gherman Titov was not only more maturely perceptive, but more sophisticated and critically discriminating. A man of strong opinions and stubborn prejudice, Titov viewed the cosmonauts' preflight activities with some of the keenness of observation of a good journalist and some of the questioning detachment of, for instance, Holden Caulfield in J. D. Salinger's deft *Catcher in the Rye*.

Titov was an excellent gymnast, but he hated to run. He often vented his spleen on one particularly cheerful, robust instructor who, from his position in the center of a circle, gleefully shouted "Run, dammit, run!" to the elite trainees under his control. Titov also secretly resented the hordes of constantly scrutinizing doctors who, often on the merest suggestion of evidence, thinned and pared the cosmonaut ranks with an abrupt and impersonal dismissal from the program. The cold steel contest, he felt, was between doctors and cosmonauts, and in the running and winning of it, he frequently lost sight of his earlier imagined role as space pilot of cloud nine. The strain of wondering who would be the next one to be eliminated irritated but did not break him.

As seen through the eyes of Gherman Titov and other cosmonauts and as described by them, their backbreaking physical training exceeded by a considerable margin the preparations of the first seven American astronauts. Likewise, there is a great amount of evidence to indicate that the technical thoroughness of the Russians' training, especially their minute preparations for possible emergencies, was and is given more man-hour application than our own. The relative emphasis on training requirements is of no great import and is somewhat academic, since the evidence is overwhelming that both groups of spacemen were superbly trained. This was impressed on me as an observer of considerable portions of American astronaut scheduling and training and as a long-range student of Soviet space activities. Evidence of the thoroughness—and, in certain respects, superiority—of the Russian training program is worth mentioning, however, if for no other reason than to forestall those crystal-balling mythmakers who sustain their reputations largely by saying and predicting precisely what they themselves and a great many other Americans would most like to hear. The Soviet space program has been subjected to so much of this popularized pontification and partisan prophecy that perhaps a few items of record are called for. It has often been said, for instance, that some Russian cosmonauts, especially Gherman Titov, became nauseated and sick in space, while our astronauts did not, because American astronauts were better-trained physically and mentally. Perhaps we had better look at the physical and technical training Titov and his fellow cosmonauts actually had.

The deliberately programmed terrors of some of the Soviet space-flight simulators could have been designed by a sadist. Of all of them, the one Titov calls "the chamber of horrors" was the impersonal granddaddy. The cosmonauts' introduction to this monster was grimly foreboding. "This is silence, comrades," said one doctor, "that can literally drive you out of your mind." Another briefing officer said: "You will see and do things here that you have never seen or done before—

and perhaps will never do again. That's up to each of you, mostly. But I assure you that in this place you will need every ounce of physical and mental fitness that you possess. No, let me rephrase that, comrades. I do not assure you of this, it is a *warning* for you. This is going to be tough."

The chamber of silence itself was a huge shell, mounted on rubber shock absorbers and suspended in the middle of a large laboratory. The walls and the single door were 16 inches thick. Two round ports, each with two layers of glass, could be sealed from the outside with a metal cover. Inside, furnishings consisted of a steel bed, table, and an exact replica of the *Vostok* contour couch and the array of *Vostok* instruments and controls.

At first the cosmonauts were taken inside as a group and the door closed. The silence, Titov reports, was absolutely oppressive and uncanny. Voices were loud and undistinguishable. "The walls," said Titov, "seemed to be bending over, moving in. Each of us wanted to get out."

So completely had all sound and vibrations been eliminated and so oppressive was the tomblike stillness that even in a group, one member, Yuri Gagarin, suddenly felt compelled to yank the door open, to the amusement of the waiting and expectant scientists who had dreamed up and fashioned the chamber.

In turn, each cosmonaut had to endure longer and longer periods of absolute silence, which seemed to crash in upon them with physical force. Titov, especially, came to dread and fear the incredible lack of sound in what he called "The mirror of truth—when you come suddenly face to face with yourself." When low waves of nausea moved through him, doctors observed the effects closely, then confined him for even longer periods, until he reached and eventually passed a stage of acute psychological discomfort. He was ordered to sit for hours, then days, in the *Vostok* seat, without motion or sound. After his burial within an avalanche of silence, he was invariably startled at the explosion of sound when he was finally ordered to open the door.

If Titov had not succeeded in overcoming his nausea, he would have been eliminated, as many cosmonauts were.

Instead, he entered a second and more rigorous phase, in which internal temperatures were drastically altered and he was confined for longer and longer periods. *Vostok* instruments were placed before him, and he was given complicated work assignments, ingenious emergencies to solve, day and night on end. Then sudden idleness was imposed for periods stretching up to three days. In his entombment, he could hear the coursing of the blood through his veins and arteries, the gurgling thump and metronomic crashing of his heart, the hoarse, wheezing sound of his own breathing. He emerged from one such eerie confinement to learn that he had been imprisoned in silence for fifteen days. As he returned to class work, he vastly enjoyed what he called the "wonderful melody of the everyday world. Every chink, bang, whimper, scratch, gasp, wheeze, knock, clang, and screech came to my ears as music!"

But Titov passed this endurance test as he also passed his diabolical confrontation with other human stress devices. He was fired out of a lumbering Ilyushin transport by a cannon shell detonated beneath his seat. He was similarly ejected from two-seater jets while in strenuous, high-G vertical banks. On delayed parachute jumps, he forced himself in the biting cold into the dreaded flat spin, and practiced landings in trees, rivers, on hillsides, in rain, snow, and strong winds, and on concrete. He was also placed in a space suit inside a coffin-like padded wooden box. After the lid was closed, the box was spun and rotated in three planes at once until nausea was again provoked and he came squarely up against the limits of his endurance. He had wires hooked up to him and was shocked, vibrated, anesthetized, and hypnotized. He was sealed in his space suit and lowered into a deep tank of water, and his body was probed, pinched, drained, strained, exercised, and extended in a great variety of ways. He could readily have said, as John Glenn once said in jest: "Fortunately there are only so many openings in the human body and only so far that doctors can explore up any one of them."

In ground school, Titov's class of cosmonauts received detailed training in bioastronautics from the most outstanding

Soviet experts in aviation medicine. They learned thoroughly the effects on the body of speed, centrifugal force, acceleration, lack of oxygen, and cosmic radiation. Titov learned to tell, almost to the instant, when bubbles began to form in his tissues and bloodstream because of decreases in atmospheric pressure. He also became expert in what happens in the dreaded impact of unpressurized human lungs and the superthin atmosphere above 50,000 feet, when the lungs refuse to admit oxygen and a man drowns in his own carbon dioxide and water vapor. "No spaceman," Titov concluded from his studies and experiences, "truly leaves the earth. He must carry vital portions of it with him."

From exhaustive lectures and dozens of annotated films of previous rocket flights, the cosmonauts became highly expert in the new science of rocketry. They became as proficient at understanding the mass ratio and pounds thrust of the long, slender blazing volcanoes of energy as, while fighter pilots, they had been familiar with a jet plane's aerodynamic lift and the function of its relative angle of attack. They studied and admired the experiences of the U.S. Air Force's Colonel John Paul Stapp as he decelerated against a water barrier from a speed of 632 miles per hour in just 1.4 seconds.

By the time Titov was introduced to his *Vostok* spacecraft, he already knew that chief designer Sergei Korolev had approved his selection as the No. 2 cosmonaut in the select ranks of what had by now been rigorously thinned to twelve men. Titov's knowledge that, continued good health permitting, he would make a second Russian flight into space coincided with his growing awe and admiration of Korolev and the "marvelous instrument" he had created with the name of *Vostok*. Like Gagarin and, indeed, most Russians who knew Korolev, Titov idolized the person he called "the single most important man in the Soviet Union's cosmic flight program—a man who walks with the giants."

The day Korolev took them for their first look at his mechanical brainchild was an exuberant one for Titov. "Before us, under the brilliant arc lights," he reported, "the conical nose sparkled and flashed. She was beautiful—overwhelmingly so!"

"She's all yours," said Korolev, laughing at their eagerness. Somehow Titov managed to beat Gagarin to *Vostok* and climbed into the cockpit ahead of him. Titov became so engrossed with the layout and simplicity of the instruments that he forgot the other cosmonauts, eagerly awaiting their turn. His proprietary absorption was finally interrupted by a rare angry shout from an impatient Yuri Gagarin.

Unlike American astronauts, Titov, Gagarin, and other cosmonauts had not, up to this point, participated in any way in either the overall design or the specific cockpit layout of *Vostok*. From this point on, however, the cosmonauts, Titov in particular, vested considerable design inputs and modifications into cockpit instruments and control devices. Even as these modifications were being made, cosmonauts, particularly the No. 1 and No. 2 cosmonauts, spent endless man-hours inside *Vostok* and inside *Vostok* simulators. As astronauts learned to touch every Mercury switch and knob blindfolded, so cosmonauts got the meticulous orientation and feel of the inside of their own space cocoon—from the black plastic control handle to spacecraft status and life-support instruments.

Most of their intense *Vostok* familiarization took place at Zhukovsky Red Air Force Base in the Moscow area, but they also traveled often to tracking stations, propulsion centers, workshops, airbases, and university laboratories and research centers. So busy was their schedule that daily calisthenics was reduced to one hour. During this period Titov, who now knew he was Gagarin's backup pilot as well as the designated pilot for the second manned *Vostok* flight, suffered, along with Tamara, a personal triumph and disaster. Their first child, Igor, was born, but after several delightful months showed signs of weakness. Titov, when he learned Igor was dying, was taken out of the cosmonaut program for several weeks. When the child died, Tamara—now grim and trancelike in her grief—was sent by train to the country.

In April of 1961, on the day of Gagarin's flight, Titov also dressed in his own elaborate paraphernalia and accompanied his roommate right up to the closing of the hatch. Titov speaks movingly of his feelings as Gagarin, "Stood entirely on his own—bearing the tremendous responsibility not only of his nation and his planet, but actually of the entire race of men. These are not empty words. We believed this to be so. I had not yet met any of the astronauts of the United States, but not for a single moment do I doubt that we could literally translate our feelings into the language of one another, and find we have long carried the same thoughts."

Titov was the last to leave the concrete and steel gantry. Then he took a bus for a short ride, went down a flight of concrete stairs through a thick steel door to the underground control center at Tyuratam. The room resembled in astonishing detail the aboveground Mercury Control at Cape Canaveral—with its banks of consoles containing red, amber, and green lights, and its radarscopes, oscillographs, tape recorders, vertical wall-to-wall charts flanked by counting lapsed-time indicators. When the time-to-launch indicator winked away the final delaying minutes and seconds, Titov at last removed his orange-red space suit and got down to his ice-blue underwear.

As the rocket thundered alive, he felt its powerful vibration deep within the earth. He heard Gagarin's slightly grunted voice as acceleration forces built up. When instruments indicated Gagarin was securely in the world's first manned orbit, the control-center scientists and technicians smiled and gave a boxer's victory handclasp above their heads, but saved the exuberant back pounding for the upcoming landing.

Titov flew with Air Force General Kamanin to the landing area, as they followed the flight's progress on the radio. They landed at an Army field near Smedlovka, where Titov fought his way through the crowds to embrace Yuri Gagarin.

"Gera, my friend," Gagarin said, "it is beyond all description. It is beautiful beyond words. To see with your own eyes . . . for the first time, Gera . . . the spherical shape of the earth."

Titov stayed in the background during the Moscow celebrations and formalities. At the proper time, of course, Gagarin gave the cosmonaut group a thorough briefing, then disappeared from their midst into his new world of abiding fame, travel, and personal appearances.

The chief thing on Titov's mind was his own flight, now moved to the front burner at Tyuratam. In his attempt at undivided concentration, he was irked by the explosion of publicity and at what he called the outside interference "that had begun to accumulate irritatingly about our small group." Such irritation, especially from intrusions of the press and *paparazzi*, was also common among our astronauts, one of whom aimed a sonic boom one night squarely at the motel and bar where waiting newsmen were drinking. And just as the American astronaut Wally Schirra later resented the fact that John Glenn was off making speeches at a time when Schirra felt he needed Glenn's advice, so, too, did Titov resent the absence of his experienced friend, Yuri Gagarin. Uppermost in Titov's mind was the certain knowledge that if all went well his scheduled flight would be the first space journey and the first Russian space flight under partial manual control of the pilot.

"The news," says Titov, "that I would orbit the earth for more than seventeen times stunned and delighted me."

His biggest worry was that the forecast of mighty storms and eruptions on the surface of the sun would coincide with his flight window.

In late July of 1961, predicted and violent solar eruptions did occur that could have propelled harmful radiation inside *Vostok*. But when the storms settled, Titov was vastly relieved to learn that the forecast for launch day, August 6, 1961, was for a calm sun. This meant that our great common solar furnace, over 90 million miles away, was predicted to remain benign during Titov's flight.

Now Titov had his own understudy in cosmonaut No. 3, a man he describes as "the calmest in an emergency that I have ever known." Together they flew from the Moscow training center to Tyuratam just a few days after *Vostok 2* arrived on the pad. They found a spacecraft modified for a longer mission by the addition of new monitoring and navigation equipment, including—as John Yardley of McDonnell Aircraft had also to redesign for Mercury—a larger port.

There were no substantial changes, however, in basic components and systems; Korolev's equipment for life support, cabin pressure, atmosphere constituents, and cabin temperature had all worked marvelously on Gagarin's flight. Additions meant, however, as Titov observed, that, "Everyone else staggered under their assignments while I had little to do."

On August 4, Titov and his backup again moved to the brown concrete and wood cottage he had occupied with Gagarin, who was now far away in Canada on a speaking tour. That evening he was unable to read and, instead, drove to the launch pad, ascended the elevator, and sat alone in his now-familiar spacecraft.

The next day, on the eve of launch, the two pilots walked with Sergei Korolev along the springy turf on the airstrip perimeter. "Above all, Titov," Korolev said, "remember that you will be the first to fly with the manual control system. We know it is effective . . . but I wish you to be cautious in your first control attempts. . . . If you are forced to re-enter somewhere else besides the assigned landing place, the manual control will mean everything to you."

Later Korolev put his hand on Titov's shoulder. "Sleep well," he advised. "Your day is going to be twenty-five hours long tomorrow."

When Titov was awakened at dawn by his physician, he immediately smelled the fresh roses that had been placed on his bureau. It was a beautiful, clear day.

After calisthenics and a breakfast from *Vostok*-type bronze tubes, the cosmonauts dressed for flight, then rode in a white-and-blue bus to the launch pad. Titov made only the briefest of speeches before he ascended the elevator and climbed into his ship. Titov had been amazed to learn that Alan Shepard had had to wait five hours in *Freedom 7* before liftoff. As in the Gagarin flight, there was no delay in the countdown, and he approached his supreme moment in a state of objective calm that became emotional only when fellow cosmonauts played the same song that had been played for Gagarin.

The countdown, now in seconds, tolled inexorably. Then Titov heard the firing command: *"Natchinay zhar!"*

The moment of highest excitement for most cosmonauts and astronauts occurs just before and during liftoff, when the marvelous alchemy of ignition turns the monstrous rocket into a shuddering giant powerful enough to thrust its fueled bulk straight up and, in the roaring process, sever the anchorage to earth and rip asunder the tough sinews of gravity. Witnesses to this dramatic blast of twentieth-century sorcery imperfectly measure this excitement by the overpowering sound and thundering vibration that seem to whirl inside your chest cavity and to resonate in the marrowed tunnels of your bones. A more precise and acceptable measurement is the pulse rate, and, fortunately, this is available from those who are closest to the fine frenzied power of the erupting nozzles. At liftoff, Alan Shepard's pulse rate rose to 139. Gus Grissom, who had not yet flown, had what Mercury physician Dr. Bill Douglas called "a normal tendency for higher readings." Grissom's shot up to 170. And John Glenn, also yet to fly in space, registered a relatively placid 110 beats a minute. The pulse of Yuri Gagarin had reached 132, and now as Gherman Titov lived his own moment of truth—his heart also pumped at exactly 132 beats per minute.

At 9 A.M. on the morning of August 6, 1961, the great rocket ascended, programmed, staged, and inserted Titov into an even more precise orbit than Gagarin's. The orbit was inclined 64° 56′ to the equator, with an orbital period of 88.6 minutes. Perigee was 110 miles; apogee, on the opposite side of the earth, 159 miles. Titov suffered several seconds of severe spatial disorientation the instant he became weightless. He felt as if he were turning in a somersault with his legs up. Then he appeared to be whirling in a "strange fog." In the sudden and unexpected chaos he experienced, he was unable to distinguish the earth from the sky or to read his instrument panel. His vision cleared, however, when he shook his head, stabilizing the vestibular system of his inner ear. Sensitive ground indicators monitoring his body had quickly noted his condition, and now he heard the urgency in the communicator's voice from Spring One, code for *Vostok* ground control.

E

"*Eagle* . . . *Eagle* . . . This is Spring One. Report immediately on your condition. Over."

Now reoriented and with clear vision, Titov reported: "This is *Eagle* . . . I feel magnificent . . . feel magnificent."

It is fortunate for the earthbound that Titov's facilities returned to their normal perceptive state, for his powers of observation, already evident in this chronicle, now combined with his innate literary sensitivity, reflected in his early and lifelong love of the works of Alexander Pushkin. Gherman Titov has produced the most acute, articulate description of the view from orbit of any of the earlier space travelers. American astronauts and, to an extent, their Russian counterparts, early in their flying careers are subjected to and develop the art of understatement. As a World War II combat pilot, I noticed that the more hazardous, violent, and dramatic was an experience in the air, the more understated and coolly objective were my fellow pilots' official and informal reports. One of the badges of recognition, in fact, of men who live close to danger—as Ernest Hemingway often demonstrated in dialogue—is the disciplined cultivation of a terse and abridged understatement. Vivid or metaphorical language hints at ornamentation of the raw fact of a deeply felt experience. While this is sometimes an admirable personal trait, it often appears to leave unsaid some of the more vivid and subjective impressions and emotions that the uninitiated hunger for. Among American astronauts, Gordon Cooper, who is every bit as laconic as the screen characterizations of Gary Cooper, epitomizes the modest virtue of understatement and the resulting vice of abridgment and incompletion, from the outsider's point of view. Yuri Gagarin, who was somewhat limited in the verbal arts of self-expression, is more like Gordon Cooper. While the ebullient and at times eloquent Gherman Titov is more like John Glenn, whose early talent for description is still remembered at New Concord High School and Muskingum College in Ohio.

Titov's descriptions of his ascent from earth and the receding picture of his planet is good imagery and, occasionally, in his feeling for color, poetic. "Like a motion picture camera that speeded up enormously," he expressed it, "the darkened side of mountains sprang magically to life as my continuing ascent and increasing velocity scattered sunlight in all directions. I stared at plowed fields that shone with a deep blue-black color . . . at lemon-colored fields of harvested wheat and at forests of deep smoke-green that seemed to float past the porthole. The clouds drifting above them were amazingly sharp and clear in their definition, like wind-filled sails hanging just above the surface. . . . The earth flashed as a million-faceted gem, an extraordinary array of vivid hues that were strangely gentle in their play across the receding surface of the world. The light streaming into the cabin carried a strange shade as though it were filtered through stained glass."

When still higher, Titov saw a picture as if magnified and intensified by a sort of narcosis of the cosmos. He speaks of "a planet enveloped in a blue coating and framed with a brilliant, radiant border" and "the terrible, intense brightness of the sun contrasting with the inky blackness of the planet's shadow with huge stars above glittering like diamonds, while a lovely, powder-blue halo surrounded the planet." The African continent was "like a strangely mottled leopard skin with green jungle sprinkled across the yellow mass." The Indian Ocean appeared "a rich indigo blue," the Gulf of Mexico "as a startling salad-green color" and the Mediterranean, the most beautiful of all, "glistened like a vast sea of shining emeralds." Rio de Janeiro was a shimmering light that "became a wonderfully rich gold dust, sparkling and gleaming against a backdrop of velvet blackness. . . . It was marvelous to ghost through space at three hundred miles every minute!"

It is clear that he brought his gaze back to his mundane instruments with some difficulty. "If space has its poetry," he observed, "it also has its prose." Most of Titov's oral descriptions to the ground contained the familar clichés that one thinks of first in moments of high excitement, television scripts notwithstanding.

During his seventeen orbits, Titov followed Korolev's calm, radioed instructions as he controlled attitude manually. Often he pointed his hand-held Konuass camera at what he called, "The most intense and incredible colors that any man had ever seen." He also took pictures of himself winking in delight and of his logbook suspended magically in the air. On his sixth orbit, he heard from the ground that he had been promoted to the rank of major and that Gagarin was flying east to greet him upon his return.

During his flight, Titov, an indifferent diner despite Tamara's excellence as a cook, ate without enthusiasm and similarly performed prescribed exercises that, nevertheless, left him feeling better. He felt a different kind of perhaps psychic ecstasy when he practiced manual control of *Vostok* for as long as twenty minutes. "I closed my right hand smoothly and surely around the black handle," he explained. "Immediately a thrill shot through me! What a tremendous feeling to manipulate with just my hand the mass of a spaceship plunging through a vacuum at nearly eighteen thousand miles per hour! . . . In the palm of my right hand was the precision control for the mightiest piece of equipment in which man has ever flown, a vessel of space that responded to the flexing of my wrist muscles and the pressure of my fingers like a well-trained animal."

After a half-dozen orbits or so, Titov was startled to obtain a sudden night visibility that penetrated the former "inky blackness" and "dark void" of the earth's shadowed side. He could now see the night clouds of earth and the bulging flanks of its nocturnal hemisphere. This led him to speculate that he was aided by the phenomenon known as "earth shine" or "air glow" that has been under investigation in the United States.

Dawns, which occurred every forty-five minutes, were to him "an explosive arrival of dazzling brilliance" while the more gradual twilights, occurring at the same interval "were an unhurried process to delight the eyes and souls of poets," and the Churchillian artist in Gherman Titov rose to the old, yet new task of capturing the sunsets of earth. "Marching across the planet," he wrote, "in a circlet of deep red-orange, the vivid sunset yielded unwillingly to the dark shrouds that advanced to envelop the light of day. Fiery colors blazed along the tidal wave of demarcation, flashing colors to the horizon where the blue halo increased its richness of hues until, by some magic of transformation, night reigned supreme."

It is a wonder, indeed, that a man of such sensitivity could survive six such sunsets within the confines of an average eight-hour workday. So elated was Titov and so caught in the grip of his new rapture of the cosmos (somewhat akin to the deep-sea diver's rapture of the deep), so overpowering was the pristine perspective that he consumed like an injected potion in all the awe and wonder of his being, that at one point in his flight he replied to a call from earth by crying, "I am Eagle! I am Eagle!"

Back beneath the sullied atmosphere and upon the hard ground, Titov said that he felt "no later reflection or regret" at having so replied to the summons of earthmen.

The remainder of Titov's seventeen revolutions, except for the drama and symbolism of his return to earth, was nearly routine. Although the severe disorientation he felt at the onset of weightlessness did not return, he nevertheless felt extreme fatigue after about the halfway point of his flight. Gradually, as the fluid in the otoliths of the inner ear lost its gravity-accustomed position, Titov again suffered spells of dizziness and nausea. "I was," he reported, "having a difficult time maintaining a sense of balance." His mental acuity was also impaired. He consciously decided to move his head as little as possible; yet he was suddenly startled to discover he was unconsciously moving his head. The action caused another wave of nausea. Finally, he became so susceptible to movement of any kind that he was affected by even the motion of dials and meters on his instrument panel. But as symptoms recurred, a calculating part of his brain observed and recorded them objectively in his notebook. "These reactions," it is important to note, "at no time interfered with the performance of my duties as they were scheduled," Titov later reported.

He was scheduled just after his seventh orbit to become the first spaceman to attempt sleep. At first, however, he was startled to wakefulness as his arms floated out in front of him, sleepwalker fashion. Finally, he learned the disturbing posture could be avoided by tucking his arms inside one of his seat belts. He found he could sleep thus with no necessity whatsoever, in his state of weightlessness, to turn his body from time to time as one does in a bed. He awoke, vaguely disturbed, once in the tenth and once in the eleventh orbit, before he settled down to something like normal sleep.

Upon awakening, he found that his required exercise brought his mind and body back to top condition, and he quickly caught himself up in his schedule of activities. Early on the earth morning of August 7, he leisurely marveled at the grandeur of his last orbital sunset. "Through the wonder, I felt there intruded the very grim facts of life," he observed. "Beautiful as was this seventeenth dawn, it was also dangerous."

Preparations for landing brought fresh excitement. He found himself longing to return to earth but admitted, "This is the moment that man in space fears." Titov had a special reason for apprehension. Whereas Gagarin had ridden in an automated re-entry all the way to earth impact inside the comfortable cocoon of *Vostok*, Titov was scheduled to test out Korolev's alternate system of landing. From the warm, safe, automatically controlled, encapsulated atmosphere, Titov was to be catapulted clear of the descending *Vostok* and land under his individual parachute. As re-entry pressures mounted, and he looked out at the fire, "like staring into the blazing maw of an erupting volcano," he was partially comforted by what Gagarin had told him: "The overloads were tremendous but just to get back to earth again I could have supported a whole mountain on my shoulders. . . . I was shouting like a drunk again and again . . . 'I'll be home! I'll be home!' "

The philosopher in Titov came through even in those moments of high stress. He later observed: "Without warning, the yearning for earth, for home, strikes hard. The time I had just spent beyond the world showed me clearly enough that there will be for men in space a monotony and a loneliness; but these are negative emotions and they cannot possibly compare with the exultation of anticipating the feel of solid earth beneath your feet . . . and the desire to see the heavens where they have always been. Up. Above me!"

His reunion was fast approaching. He watched the sweep hand on the chronometer approach zero. At zero minus three seconds, a red light came on. Then he heard thunder crash in his ears as he and his entire seat were explosively plunged from Vostok. A few seconds later, he felt his seat separate and his own chute "opened with a wonderful, muffled boom above me." Below him, drifting earthward on its white parachute, was the now-ruptured Vostok that had been his space home for some twenty-five hours. His landing, like Gagarin's, was close to target center. Below him, he could already see small figures running across the ground.

He plunked down into a plowed field, and in his exhilaration he sat up laughing and rubbed the brown earth gleefully against his face. Then, as two men on a motorcycle bounced crazily toward him, he stood up and held both his arms aloft.

"Are you Titov?" the men shouted.

"The same," he said.

Titov, as had Gagarin before him, went through the riotous receptions, the calm and penetrating debriefings, the grand ceremonial welcome to Moscow, the lonely and symbolic walk down the long red carpet, and the joy of reunion with his wife.

Then one quiet clear night he and Gagarin, the only two Russians to orbit the earth, walked along a riverbank together. The subject they discussed, Titov relates, was space, but a more distant part than either of them had seen. As they walked under the vault of bright stars, the two spacemen talked about Mars.

10
Vostok and Its Pilots

Details on *Vostok* construction and components and how they compare with those of the U.S. counterpart, the Mercury spacecraft, have been omitted until now primarily because *Vostok*, like Mercury, was in a fairly constant state of evolution. The last of this series, *Vostok 6*, flown by Valentina Tereshkova, for instance, was different from *Vostok 1* flown by Gagarin more than three years earlier.

The U.S. at the time knew nothing about *Vostok* and learned very little from the initial and elliptic Soviet announcements. My first view of Korolev's marvelous machine presented me with an engineering enigma, since all I had studied by way of comparison were the only two other spacecraft then on the face of the earth, Mercury and Gemini. Fortunately, I had often talked with their designer, John Yardley of McDonnell Aircraft, and I had sat long minutes in Mercury and had once examined it and its still hot thrusters while it was still wet from an aborted flight at Cape Kennedy. I knew and had often written of details of Gemini's construction. But *Vostok*, at first, was a puzzler. It was only after I broke her subsystems down one at a time and made the rather obvious comparison to the subsystems and components of U.S. spacecraft that I came to appreciate some of Korolev's engineering genius as well, I might add, as that of John Yardley.

It is remarkable in how many engineering aspects the Soviet space program parallels that of the U.S. The basic design specifications, for instance, facing Sergei Korolev and his team were essentially the same as those that faced John Yardley and his colleagues in St. Louis. Both had the unprecedented task of constructing what was essentially a hollow meteor that would not be ruptured by zero pressure, subzero cold, extreme heat, the merciless buffeting of ever-denser molecules of air, and, finally, impact on land or water. Both, in order to keep a man alive, had to create and sustain, in addition, a viable atmosphere. Each had to have a rudimentary control system that would maintain a desired attitude—both manually and automatically. Each had to know its own speed, height, and direction of flight in relation to the earth and a projected impact point. Each needed pyrotechnic devices to brake precisely its orbital momentum, and each needed drogue and descent parachutes and space-pilot escape modes. Each also had to contain communication, radiation-shielding, and data-gathering equipment. Finally, each spacecraft had to have adequate room for food, fuel, atmosphere components, and backup systems. The wonder is, given virtually identical criteria, that each ended up so different and so successful. This is particularly remarkable when we realize that, at the level of *applied* research, there was virtually no interchange between the two sets of designers. The biggest difference facing the insulated teams of designers was the difference in available booster power. Because of the relatively low lifting power of the Mercury-Atlas booster (approximately 360,000 pounds thrust) Mercury, fully loaded, could not weigh more than one ton. The Vostoks, on the other hand, with their more powerful boosters, could weigh an average of 10,428 pounds, nearly five tons. Much of the Russians' available weight, however, went into the extremely heavy outer shell of Vostok. It is not known what this shell alone weighed, but it appeared upon examination to be about two inches of solid steel alloy. Reentry heat insulation was provided by extremely thin sheets of reflective, aluminum-like metal plus layers of an asbestos-like material crammed into a honeycomb of hexagonal cells. "Sharik" (meaning sphere), as the Russians called *Vostok*, had three hatches—the parachute hatch, the technological hatch, and the round exit hatch. In addition, there were three shuttered ports, all of refractory glass. The part nearest the pilot had a Yzer optical device affixed to it to assist the cosmonaut in manually operating his ship. This direct and simple design solution, whose spherical product suggests the boiler of a locomotive sheathed in aluminum and asbestos, is in marked contrast to the frame and skin of Mercury. Designer John Yardley called for an outer shell of nickel-cobalt alloy, thinner than a dime but corrugated for added strength. The inner shell of lightweight titanium was separated from the outer shell by an inch and a half of insulation. Mercury's leading and re-entry surface, however, received special treatment: it was protected on suborbital flights by the largest piece of beryllium ever forged, a six-foot disk known as a heat shield. On orbital flights, an even tougher resin and glass-fiber heat shield protected Mercury's interior by utilizing the ablation—or surface boiling—principle to dissipate re-entry heat.

Mercury's interior was jammed with seven miles of wire and over 136 instruments, toggles, or switches, most of which were concentrated just above the pilot's face. Directly in front of the Vostok's pilot, however, were just eight instruments, including a television camera. Located to the left of the cosmonaut were the parachute container, drinking water, the ship's control panel, the backup heat regulation system, the loading system direction finder, and a tape recorder. A few other dials and color indicators were located on either side, along with a radio and old-fashioned telegraph key. Like Mercury, Vostok had a flight-path positioning instrument, a rotating globe of the planet about five inches in diameter. Vostok also had a unique speedometer that registered as white figures the number of revolutions and decimal points of revolution around the earth. The ship's control handle—for yaw, pitch, and roll movements—resembled its Mercury counterpart. It was a polished, black plastic handle with grooves for each of the fingers of the right hand and a small white button like a jet pilot's trim button—beneath the thumb. One major difference and a number of minor differences between Vostok and Mercury had not been part of Sergei Korolev's original plans. In Project Mercury, U.S. astronauts were from the beginning assigned specific areas of the spacecraft construction and invited to suggest pilot-oriented modifications. John Glenn, for instance, was assigned to cockpit layout; Gus Grissom got the attitude-control system; Alan Shepard was assigned to recovery techniques. But Russian cosmonauts did not even see Vostok until it was virtually complete. Several of them silently noted things that they felt needed modification from the pilot's viewpoint, but so universal were the awe and admiration for the chief spacecraft designer—or "chief constructor," as cosmonauts called him—that the future space pilots were at first reluctant to make their suggestions known. Korolev was known to be a very warm and human person, but there existed about him, nevertheless, a commanding aura of authority.

His very anonymity suggested the mystique of the cosmic sorcerer and contributed to a certain insular rank. Although the Russian press frequently quoted his von Braunish prophecies about space and Radio Moscow sometimes beamed his actual voice across the vastness of Russia, he was never identified by name. This shroud of mystery lent a certain enchantment, and his presence was known to command any gathering of his co-workers, whether or not he was actually doing the briefing. Nevertheless, after the class of twelve cosmonauts was allowed to sit in his cosmic creation and began to attend classes on its inner workings, a number of cosmonauts began to raise questions privately about the strange and marvelous vehicle assigned to them. Gherman Titov, in particular, thought a number of things could be better designed and modified. He had the proper amount of audacity—if not downright brashness—to detail his objections on a set of drawings and lay them before the chief constructor.

Korolev studied them silently for ten minutes, while Titov stared at the wall and shifted his weight from one foot to another. Finally the chief designer leaned back in his chair. " 'You realize, Titov,' he said, 'that if I were to follow your recommendations it would mean just about rebuilding the entire spaceship?'—I recall that I did ask for questions and suggestions, but I never expected anything like this." When

Titov returned without having been struck by lightning, the other cosmonauts were encouraged to submit suggestions of their own. They flooded Korolev's office with drawings. In a subsequent and solemn meeting, when Korolev "was not in one of his better moods," as one of them put it, the chief designer dismissed most suggestions one by one with a brilliant discussion of the basic principles that had guided him in designing the first manned spacecraft. He patiently explained the difference between a seemingly simple solution and the hard realities of space-flight engineering. When the meeting was over, Titov was relieved to notice, however, that some of his suggestions had not only not been criticized, they had not even been mentioned. But it was not until later that Titov learned the outcome of his audacity. He was summoned to Korolev, who privately escorted him to the assembly hangar, where a group of exhausted men stood around a somewhat new Vostok.

"There she is, Titov," Korolev cried; "she is finished. Now, tell me, do you notice anything different about our space vessel?"

The outside seemed the same, but when Titov lowered himself inside he found a changed cockpit. Some controls he had criticized were now easier to reach. Panels that he had pointed out were too dim and small to read now glowed clearly at the flick of a switch. The most important change, however, had not been suggested by a cosmonaut. The sphere of Vostok is basically two compartments—one for the pilot and a larger one for equipment and experiments. In the first version, the pilot's contour couch had been fixed and rigid in relation to the frame of the spacecraft, as in Mercury. Now, however, Titov was amazed to see that the entire couch and seat assembly in the pilot's compartment had been remounted on a system of rings and bearings. It could actually turn inside the spacecraft in a complete circle. This, it was immediately apparent, would enable the pilot to maintain his own horizontal attitude in relation to the earth's surface. The device also would enable him to position himself in relation to sun glare and his most desired view outside his glass ports. This was an advantage Mercury space pilots—whose vision was sometimes impaired by a fixed and penetrating shaft of brilliant sunlight—did not enjoy. These Vostok changes, which were made prior to Gagarin's flight, were highly approved by the cosmonauts.

Vostok was simpler than Mercury, but it was by no means crude. Alekse Ivanov recalls from his notes that Vostok contained 241 electric lamps, over 6,000 transistors, 56 electric motors, and about 800 relays and switches.

One of the most significant differences between Vostok and Mercury was in its atmosphere constituents—differences widely realized for the first time following the fatal oxygen fire in the Apollo spacecraft on January 26, 1967. In Mercury, Gemini, and Apollo, astronauts subsisted on 100 per cent oxygen pressurized to about one-third sea-level atmospheric pressure, about 5 pounds per square inch. Vostok, however, like the U.S. Air Force's Manned Orbital Laboratory, was designed to use a two- or three-gas system consisting of about 24 per cent oxygen plus other gases pressurized at or slightly higher than normal sea-level pressure—about 15 pounds per square inch. The other gases present in Vostok were nitrogen, possibly some helium, and traces of carbon dioxide.

The atmospheric choices facing closed-container designers is an interesting problem. The air we breathe on earth is roughly 20 per cent oxygen and 80 per cent nitrogen, but nitrogen production in a closed system is expensive in weight and space. Oxygen readily supports life, but, negatively, it also supports hazardous combustion and in its pure state, often causes such adverse side effects as eye and throat irritation, hearing impairment, and a clogged chest. The higher the pressure the more these effects are magnified until, finally, under great pressure oxygen becomes lethal. Nitrogen, too, becomes lethal under high pressure. That is why when astronaut Scott Carpenter decided to live beneath the Pacific Ocean in a closed container known as *Sea Lab 2*, his oxygen content was reduced to 4 per cent, nitrogen to 11 per cent, and helium had to become the principal constituent (85 per cent), because it does not become narcotic at the seven-times-normal pressure required by Carpenter's 205-foot depth.

But helium in such large amounts has disadvantages also, one of which is severe impairment of hearing. The biggest disadvantage of pure oxygen in a closed container is the violent way it supports combustion, as demonstrated in the tragic Apollo fire. The biggest advantages are that, by excluding nitrogen, possible malfunctions of highly complex nitrogen generators and atmospheric controls are eliminated and—more important to Mercury designers—the rupture of the spacecraft by a meteorite would not induce a possibly fatal attack of the bends. If a nitrogen-rich earth-like atmosphere is punctured and vacated in the vacuum of space, the rapid decompression could cause nitrogen dissolved in a space pilot's blood to come suddenly out of solution, forming fatal gas bubbles in tissues. That is why Korolev had to design an airlock in Vostok's successor, Voskhod (which also used a mixed-gas system), before cosmonaut Colonel Alexei Leonov could venture outside his spacecraft for the first walk in space. American astronaut Ed White could depressurize his Gemini spacecraft and go directly into a vacuum because he was breathing pure oxygen and had little nitrogen in his blood. He, therefore, did not risk fatal bends. Korolev avoided the bends problem for Leonov by having him first rid himself of the spacecraft's nitrogen by temporarily breathing pure oxygen. Then Leonov entered the airlock, sealed his suit, lowered the pressure, and stepped out into space still breathing pure oxygen.

When Korolev was faced with a choice of atmospheres for Vostok, Russian investigation of the meteorite hazard was slightly more advanced than our own. In a calculated risk, he decided to accept the hazard of possible sudden decompression of a part-nitrogen atmosphere and to design a very simple air-regenerative system to reduce the probability of malfunctions. In making the decision, he automatically rejected the antithetical hazard of a violent pure-oxygen fire. What he turned out was a noisy air-regeneration contraption that worked, nevertheless, like a reliable donkey engine. The chemical devices that generated oxygen were located behind and to the right of the pilot's seat. The absorption of carbon dioxide, water vapors, and the generation of gases was an automatic process controlled by sensors and special chemical compounds. Korolev linked the regenerative system to an air-conditioning system that furnished the 24 per cent oxygen atmosphere, maintained normal pressure of about 750 mm. of mercury, and a relative humidity of about 65 per cent. An automatic heat-adjusting system used both an air circuit and a liquid circuit, which converged in a heat-exchange unit in the ship's cabin. Once the pilot adjusted the temperature (normally about 19° C), the system automatically governed and retained it.

Perhaps the versatility and adaptability to space flight that Korolev's team put into Vostok is best demostrated by following it on a typical orbital flight. As it sat on the Tyuratam pad atop its rocket train, the sphere of Vostok could not actually be seen, nor could the cosmonaut inside see out, as Mercury pilots could while waiting out launch. This was because the Russians sheathed "Sharik" in a bullet-shaped, red-tipped ballistic cap for the buffeting ride upward through the atmosphere. After liftoff, however, while the first-stage booster was still thundering out its flaming gases, the space pilot pressed a button near his armrest. Immediately, an electric motor slid back a covering shield over the port nearest the cosmonaut containing the Vzor optical alignment device. During vertical ascent, green lights on the instrument panel indicated the proper functioning of the accelerometer, altimeter, and the life-support system, as well as the sequence functioning of the automatic pilot during, in turn, first-stage burnout and second stage and then third-stage ignition. Lights also indicated when the covering booster cap was jettisoned.

After the rocket assembly had curved over into a nearly horizontal position and the orbital Vostok was still attached to its third stage, the total weight and on-board fuel was 6.17 tons. The combined length was 7.35 meters. At this point, the assembly looks something like a silver tangerine nestled on a necklace of grapes (air balloons) sitting on top of a beer can with a single nozzle aft. When the cockpit chronometer hand reached the predicated point of orbit (17,750 mph) the final stage separated with a slight jolt. Vostok was now in orbit, with its choker of fifteen air balloons, instrument compartment, its exterior orientation system, micro-engines, and five antennas. While in orbit, the cosmonaut communicated to earth with Vostok's three radio-telephone systems: VHF (143.625 megacycles) for use over the Soviet Union, and two shortwave (HF) systems (9.019 and 20.006 megacycles) for longer-range use as ionospheric interference permitted. Simultaneously, a battery of transmitters relayed to the ground telemetry data on biomedical sensors and system performance, along with the television image of the cosmonaut and portions of the cockpit. (It was somewhat later before the U.S. utilized live cockpit television as a data-communications instrument.)

The life-support system was monitored by electronic watchdogs that automatically maintained oxygen concentration at 24.33 per cent of the cabin pressure and carbon-dioxide levels below 1 per cent. When the cosmonaut wished to control Vostok manually, he flipped a switch on the left side of his instrument panel and manipulated the control handle by his right knee, much in the manner of Mercury pilots. The plastic control handle activated the exterior micro-engines in the desired sequence and at the desired thrust level. Automatic functioning of this orientation system employed sensitive gyroscopic and optical elements and logical devices. Vostok had two relatively minor in-flight luxuries that Mercury did not have. If the dazzling light of the naked and unfiltered sun became a problem to the cosmonaut, he could flip a switch and electric motors whined as an opaque shield slid over the viewport. In addition, the rotating earth globe in front of him not only traced his position above the planet, it could also quickly adjust itself forward to show visually where his impact point would be if he chose to fire his retro-rocket. The early and crude development of this impact-point display in the cockpit could prove to be the prototype of some appalling orbital bombsight of the future.

The principles governing Vostok's re-entry and landing were, of course, the same as for Mercury, but there were important procedural differences. Mercury retros were fired over the Pacific as the spacecraft, inclined to an angle of 34°, approached the U.S. West Coast. The spacecraft's main 63-foot parachute blossomed out in the lower strata of the atmosphere some thirty minutes later, about 250 miles southeast of Cape Kennedy. The landing, of course, was on water. Vostok, orbiting on its inclination of approximately 65°, fired its retros over Africa, not far from Mt. Kilimanjaro, in order to land, also about half an hour later, just east of the Ural Mountains about 700 miles due north of Tyuratam. The nearest runway, at Sverdlovsk, was 880 miles east of Moscow. To align itself for re-entry, Vostok was automatically positioned by jets that got their alignment from the sun. The jets maintained the retro package forward and the cosmonaut aft. In case the braking system failed, Vostok had life support long enough (usually ten days) to permit normal orbital decay and landing by parachute. Like Mercury, its retro-rockets were fired, either automatically or manually, in sequence. After the instrument compartment had separated, the spherical "Sharik" oscillated and sometimes rotated as it slammed down through the atmosphere, although it normally soon stabilized. Because of possible rotation, however, its heat shielding was not limited to a designated and leading re-entry surface, as in Mercury, but was distributed all around the sphere. When its velocity slowed to subsonic, the automatic landing sequencer went into its own countdown as shown on the pilot's chronometer. Normally, the cover of the entrance hatch automatically opened and, two seconds later, the pilot and his seat were explosively ejected from the spacecraft. His parachute opened immediately. The pilot in his pressurized suit and still in his heavy seat continued to descend. Near the ground, his seat was separated and allowed to fall to earth. Still attached to the pilot's chute was a duffel bag containing a few basic-survival items, including emergency rations and a small orange dinghy that was triggered to inflate itself automatically if it landed in a lake or stream. As the pilot descended, his empty spacecraft popped out its own braking chute and, finally, its main chute at an altitude under 5,000 feet.

By U.S. standards, the low altitudes at which final landing chutes are deployed represent extremely low parameters—parameters that are only feasible over virtually flat topography. The presence af the towering Ural Mountains just a few miles west of the landing area and the ever-present probability of an emergency landing in unknown terrain underscores the bold nature of this delayed-chute deployment. Its advantage, of course, is that it permits a more pinpoint landing and more precise trajectory data. I was aware that Soviet landing accuracy exceeded our own, but was not prepared for a motion-picture scene Walter Cronkite and I witnessed one day in the Soviet Embassy in Washington. Earlier, a Soviet Embassy attaché, Boris Romanov, had secured for me some new Russian color film that showed a portion of the Gagarin landing sequence. I had been surprised then that a camera could have been prearranged to be on the Leninsky Put Collective Farm outside Engels. But after I called Cronkite in New York and he flew down to look at the film with me, I was even more surprised, upon more careful viewing, to spot a permanent-type wind tee in the far right corner of one of the shots showing the parachute drifting toward earth. The conclusion was inescapable: if Soviet cosmonauts could land near a preset wind tee, their re-entry accuracy was little short of amazing.

After Gherman Titov's seventeen orbits in August of 1961, the Soviets waited a year and five days before they sent an improved manned Vostok into space. At first, the surprise the Soviets had in store was not apparent. During the two days just prior to the flight—on August 9 and 10—Moscow was rife with rumors of an upcoming space spectacular. The first Soviet announcement of 1 P.M., August 11 (70 minutes after launch) that *Vostok 3* was in orbit with a single cosmonaut seemed initially only to offer the possibility that Soviet goals were simply to surpass Titov's record time in space. Selected as the pilot of *Vostok 3* was Titov's backup cosmonaut, a 32-year-old bachelor whom Russians refer to as a Chuvash, meaning that he comes from Bulgaric people who used the Turkish language. Andrian Grigoryevich Nikolayev's home village, Shorshaly, is in the black-earth region of the middle

Volga Basin—about 700 miles east of Moscow. One of six children, Nikolayev recalls that when he was eight he climbed a tall tree and announced to a group of boys on the ground that he was going to fly down. His friends dissuaded him with some difficulty. He at first studied to be a doctor but switched to forestry under the influence of his elder brother, Ivan, who worked in a timber-procurement district. Nikolayev started his career as a lumberjack in the birch and fir forests around Karelio. As he worked himself up to foreman, he also read a great deal, and his studious habits later served him during his Red Air Force training as a radio operator, a gunner, and finally a student pilot. He received his jet pilot's wings in 1954.

When he was selected for cosmonaut training in 1960, his rugged build and mass of thick brown hair attracted the attentions of a girl friend. But due to the secrecy and isolation of his cosmonaut training he was unable to tell her what the duties were that kept them apart. He asked her to wait for him, but she grew tired of waiting and married someone else. It was, perhaps, a blessing in disguise, for Nikolayev soon shared part of his preparations for space flight with the first woman selected for cosmonaut training, 26-year-old Valentina Tereshkova, a pretty, dimpled, cheerful, and spunky girl from Yaroslavl. When Valentina first checked in—along with the rest of the group of girl cosmonaut candidates—at Tyuratam's "cosmonaut hostel," it was Andrian Nikolayev who carried her bag up to her room for her. After his flight and sudden fame, Moscow newspapers called Nikolayev the "most eligible Soviet bachelor" and published pictures of him clumsily frying eggs in his bachelor apartment. He was swamped by hundreds of letters from eager Russian women, but it was soon apparent that he had eyes only for Valentina. Following the marriage on November 3, 1963, of what Russians were fond of calling "the star-crossed lovers," Valentina gave birth to a 6-pound 13-ounce daughter on June 7.

Dr. Eugene Konecci of the President's Space Council once teased a group of Russian scientists that the Nikolayev-Tereshkova marriage was "the first state-ordained wedding for science." The Russians broke out in laughter, pointed to the wedding date, the birth date of the daughter, Yelena, and finally, the weight of the baby. "Not state-ordained," one of the Russians chuckled. "Love intervened!" Actually, before their formal wedding in Moscow's "Palace of Marriages," the two space pilots had been living together in the common-law marriage practiced in the Soviet Union.

But love, marriage, and fatherhood were not the paramount things on bachelor Nikolayev's mind as he prepared for his role as cosmonaut No. 3 in the summer of 1962. The Russians had sent up only seven unmanned satellites since Titov's flight, while the U.S. had orbited John Glenn and Scott Carpenter. Nikolayev was well aware of the import of his mission in space. And as John Glenn found out, delays can either break a man's keenness or hone it to a fine edge. Nikolayev, apparently, was the kind of man who, like John Glenn, was sharpened under pressure. Titov reported in his remarks about the chamber of silence and horrors that Nikolayev had astounded everyone by emerging buoyant and cheerful after a stint inside nearly twice as long as anyone else had endured. Titov had barely managed to hold on to his faculties during a 48-hour confinement, but Nikolayev had painted pictures of considerable merit even while bathed in sweat. As the heat was increased doctors were astonished at his ability to maintain his body-temperature level psychologically for an amazing total of 96 hours.

When he emerged, smiling, with his sketches, the astonished doctors involved in the unusual test vied with each other for the privilege of owning the drawings of birch trees that had helped make possible Nikolayev's remarkable composure under stress. Titov, who became the closest of friends with his backup cosmonaut, refers to Nikolayev as "the calmest man in an emergency that I have ever known, and his solid-rock look upon the worst of disasters (even when he is personally involved) is a subject of much discussion among the scientists and doctors. Once he flew a powerful jet fighter—damaged and flaming—to a safe forced landing without power, when it seemed absolutely impossible that he would ever survive. He walked away without a scratch, already noting down on a pad in the most precise engineering

VOSTOK AND ITS PILOTS   137

terms, what had gone wrong with the machine. He speaks so seldom that when he does talk, we all listen to every word—invariably his few sentences result from careful, long thinking."

The time Nikolayev showed the most excitement was not on his own flight but when he tried to sleep in Yuri Gagarin's bed next to Titov the night before Titov's flight. Titov recalls only one sentence his tight-lipped friend said to him on the morning of the second manned flight: "You've got just the right conditions—wonderful." But during Nikolayev's own flight, launched from the same pad that lofted Titov, the quiet voice of *Sokol* (or *Falcon*) as he was called in radio sign, spoke with the privileged and freed tongue of the ethereal space farer. His voice not only came in loud and clear to Soviet tracking stations and instrumental trackers at sea—some very close to the U.S. East Coast—but was also heard in London, Tokyo, at RCA in New York, and Bochum, Germany, on the announced Soviet frequencies. Despite such evidence, there was still a large body of American man-in-the-street opinion—echoed by a few inveterately blinkered men of some influence—that manned Soviet space flights were an elaborately engineered hoax. As the reticent Nikolayev spun on and on, speculation mounted on the Russians' ultimate objective. Despite Soviet security on activities at Tyuratam, the rumors that were now spreading from Moscow to the rest of the world were essentially well-founded; the most prevalent speculation was that Nikolayev's objective was to surpass Titov's 25 hours substantially and that another cosmonaut would shortly join him in space.

As *Sokol* continued to circumscribe the planet every 88 minutes, the growing belief that he was scheduled for a long-duration flight was bolstered six hours after launch when the image of the cosmonaut sleeping in *Vostok* suddenly flashed live on Russian television screens. Then he appeared to come to life, to move his arms and tend his instruments. Korolev had installed in *Vostok 3* two television monitors—one profile, one full-face—that transmitted ten viewing frames per second to form a 400-line picture (compared to the commercial U.S. standard of 30 frames per second and 525 lines). Later, Nikolayev was seen to smile on television as he talked to Khrushchev, who remarked mysteriously that he would have to make "many more orbits." In the early phase of his flight, Nikolayev also became the first spaceman to eat "natural" food.

While he was still alone in orbit, Russia tried to register an additional propaganda gambit by handing the U.S. Embassy in Moscow a somewhat loaded document. "The United States," said the curious injunction, "must refrain from carrying out any measures which could in any degree hinder the exploration of outer space for peaceful purposes or endanger the cosmonaut's life." The Soviets were referring, of course, to high-altitude nuclear tests, which Russia had previously criticized (following our nuclear test a month earlier) as creating "conditions dangerous to a cosmonaut's life and health." But the U.S. had scheduled no such tests within the next several weeks and took some of the air out of a loaded balloon by responding immediately that it in no way planned to interfere with the flight and that it wished Nikolayev "a safe flight and a happy landing."

Among the things the Soviets had not announced was their deep concern, following Titov's nausea and special disorientation, over man's adaptability to space flight. Medical sensors on Nikolayev and his verbal reports to the ground early established that he had not experienced a repetition of Titov's confusion and dizziness at the onslaught of the weightless state. The cosmonaut physician Dr. Oleg Gazenko told me that this evidence figured initially in their gradually evolving conclusion that Titov's difficulty was due to his peculiar, individual, and somewhat isolated response to a situation that basically could not be duplicated on earth. Nikolayev continued to register and report his excellent condition as he approached Titov's record seventeen orbits. When the taciturn spaceman streaked in over the Soviet Union from Eastern Europe on his sixteenth orbit, he was very precisely tracked at Soviet stations near Yevpatory in the Crimea and at Ashkhabad, just over the border from Iran near the old Samarkand Trail of Genghis Khan. When Tyuratam and its secret underground control link had an exact projection of his pass over Kazakhstan, Korolev started the automatic launch sequences on *Vostok 4*. This initiated the world's first venture into the subtler areas of the new science of orbital mechanics.

In order for *Vostok 4* to end up with at least part of its orbit near *Vostok 3*, its precise time of launch, its launch inclination, and its cut-off velocity had to be realized within extremely fine tolerances—tolerances that ultimately could permit such intricate conjunctions as rendezvous, inspection of a hostile orbiting vehicle, assembly and resupply of permanent space stations, and eventually, destructive interception or capture of satellite ordnance or surveillance and communications relay platforms. So it was to be a significant pioneering attempt complicated by a somewhat lopsided or elliptical orbit. Nikolayev's apogee was 157 miles, his perigee 111 miles. It was also complicated by the speed of the earth's rotation, since the initiation of the rendezvous was based on calculations of the speed at which the earth rotated the launch pad approximately underneath the *Vostok 3* orbit. This would permit both spacecraft to end up in the same orbital plane. Otherwise, they would spin like two hoops differently inclined. The rare and antic confluence of the seam lines of such spinning loops—representing the two spacecraft—suggests the inordinate difficulty of bringing about, or even calculating, their rare periods of proximity. The two orbits also had to come close to averaging the same distance above the earth. Otherwise the speed of the inside orbiting Vostok, in order to maintain the gravity balance of its centrifugal force, would be faster than that of the outside orbiting Vostok. And again, confluence would be vastly complicated.

The original orbit of *Vostok 3* had been based on Soviet estimates of the position and quantity of the now stratified radiation resulting from the U.S. high-altitude nuclear test thirty-two days earlier on July 9. The Soviets claimed their flights were delayed, in fact, until the residual radiation reached "permissible levels at the altitude the ships were flying." But, to be doubly cautious, they had put Nikolayev into an orbit that provided him minimum radiation intake from the energy trapped in the earth's radiation belts, and they set up special instruments to measure accurately his periodic exposure to the higher millirad counts.

Vostok 4, therefore, was aimed for the same orbital plane and same elliptical orbit that computers showed *Vostok 3* to be in. Orbital adjustments from each spacecraft then available to the two cosmonauts were extremely rudimentary and entirely insufficient to evoke measurable changes in orbital plane. So the burden of accuracy was upon the precision of launch and insertion.

At fourteen seconds after 11:02 A.M. on Sunday, August 12, *Vostok 4*, on the same pad from which *Vostok 3* had been launched, roared aloft bearing a 32-year-old former Ukrainnian shepherd, Pavel Romanovich Popovich. Later that day, the Russians announced only that Nikolayev and Popovich had come within sight of each other. But, as is usual on Soviet space flights, various and publicity-conscious international sources attempted to fill in this and other data. Japanese sources reported they were within 75 miles of each other at their closest point; U.S. sources, slipping a decimal, reported at one point 796 miles. A Danish Communist newspaper reported each Vostok weighed 8.5 tons. England's talkative director of its Jodrell Bank radio observatory, Sir Bernard Lovell, had by now grown fond of the transatlantic calls placed to him by American science editors and already had managed to create the impression that he alone knew what the Russians were up to. After all he controlled a superb tracking instrument and had once been invited by the Soviets to inspect some of their tracking stations. News-hungry editors often tend to equate the validity of Sir Bernard's remarks to the length and cost of the radio-telephone calls placed to him between editions. Sir Bernard is on safe ground when he reports what his instruments read, but he often cannot resist converting trajectory data into prophecy, which, he early discovered, is also quite readily quotable. He expansively told Reuters, first of all, that the flight of the two spaceships was "the most remarkable development man has ever seen." Lovell went on to reveal that the two ships might link up, but their most important mission was to further test communications and navigation in space. He also authoritatively volunteered that there had been twelve secret American space launches so far in 1962 and that there were grounds for believing that the U.S. might already have attempted similar space maneuvers. This would have been news, indeed, to Wally Schirra, who at Cape Canaveral was then preparing to orbit the earth six times without the company of anything—except perhaps John Glenn's "fireflies."

With *Vostok 3* and *4*—each of which weighed about five tons—Russia came remarkably close to duplicate orbits. Popovich's initial orbit was inclined slightly from Nikolayev's by just two minutes, or one-thirtieth of a degree. So the two orbital hoops did not quite spin in the same plane. Popovich's ellipsis also nearly coincided with Nikolayev's. On the initial orbit, at apogee, he rose just about two miles higher and on perigee dipped about two miles lower. This meant that Nikolayev, on his slightly smaller orbit, was moving faster through space in relation to Popovich and, consequently, was steadily lengthening the gap between them from their closest confluence—about four miles—to their farthest confluence of about 1,860 miles. According to the seemingly inverted and paradoxical laws of orbital mechanics, Popovich would have to slow down to catch a smaller orbit. Had Russia then had a reliable on-board computer and the capability to measure and make the necessary corrective burn, so close were the two orbits initially that the burn would have required a negligible fuel weight.

From inside *Vostok 4*, Russians heard the unfamiliar accents of a new spaceman speaking under the call sign of *Berkut*, or *Golden Eagle*. His more jovial personality and infectious gaiety brightened the orbital world of the more self-contained and reticent Nikolayev as they chatted back and forth.

"I am Golden Eagle calling Falcon," Popovich said at one point, after a seven-hour sleep. "Am in perfect mood. Slept well. Feeling marvelous." When he appeared on one seven-minute take on Soviet television screens, he grinned and clowned by tossing and floating a small object about in his cabin. Russians avidly read the released biographical credentials of their newest instant hero. As a youth, Pavel Romanovich Popovich prided himself on not wearing shoes on his toughened herdsman's feet until the first snows fell. Like Gagarin's home town, Popovich's village of Uzin near Kiev in western Russia had early been overrun by the Germans, actually when Pavel was eleven. He studied in vocational schools near Kiev, then moved east to attend a technical school in the Ural Mountains. In 1951, the year he graduated as a building technician, he also met his future wife, Marina, at her graduation ball. They were married four years later. Popovich joined the Red Army, became an Air Force pilot, and received the Order of the Red Star for an undisclosed "government assignment" in the Soviet Arctic. After his selection for cosmonaut training, he exercised his excellent tenor voice while in the isolation chamber by singing operatic arias and folk songs for hours at a time. One of the physicians who appreciated his voice was Dr. Boris Yegorov, who himself later became a space pilot.

Popovich's wife, Marina Lavrentyerna, like Gordon Cooper's wife, Trudy, is an experienced and capable pilot. Marina is a green-eyed Siberian blonde, and at the time of her husband's space flight, she was credited with more flying hours than he was. She was a stunt pilot in prop planes, a competitive speed flyer in jets, and a helicopter pilot.

As Popovich and Nikolayev spun on and on in space, the Russian "Shoot now, talk later" policy again worked in their favor. As the space-twin denouement remained in suspension, the world's front pages were assured. The Chicago *Daily News* Paris correspondent, Paul Ghali, attempted to analyze it. "Virtually every French paper," he observed, "has its scientific explanation of what the Russians were attempting to do. Had this been an American feat, many people in Paris would have observed a maze of heavy technical information that would have been issued immediately. The Russian technique, on the contrary, is to withhold essential details and to provoke so much speculation that they get fuller news coverage." By Tuesday, August 14, the world press chronicled the fact that Nikolayev had become the first man to travel more than a million miles in space, more than enough for two trips to the moon and a distance it would take a modern jetliner two and a half months to fly. His and Popovich's excellent condition and their ability to eat a hearty space banquet of cutlets, roast veal, chicken fillets, and pies and juices had already resolved Soviet apprehensions about man's adaptability to long space flights.

It soon became apparent to the world that Nikolayev and Popovich did not intend to link up in space because their Vostoks were not programmed for what U.S. engineers called "the transfer maneuver," the final and delicate insertion of one spacecraft into the same plane and same orbit as another spacecraft, in close proximity to it, and with equal velocity. After Nikolayev had orbited the earth 64 times in approximately four days, the two cosmonauts descended to earth, like Gherman Titov, on their individual parachutes. They touched ground six minutes apart in the hill and desert country south of Karaganda, a largely Spanish-settled city 1,500 miles southeast of Moscow. During the long flight that carried Nikolayev 1.6 million miles, the Soviet people had built themselves up to a high pitch of emotion.

As the newest spacemen began their rounds of celebrations, two revealing and antithetical statements were made by spokesmen of East and West. In Moscow, Professor Petrovich wrote enthusiastically: "The '60s will witness a flight to the moon. There can be no doubt that in the '70s man will visit Venus and Mars." And in London, former President Eisenhower held a press conference in which he said there was no gap between the space programs of Russia and the U.S.

At the press conference of the cosmonauts, Nikolayev soberly and somewhat grimly read his prepared speech. Popovich read a prepared speech also, but his manner was more open and jovial. The most remarkable thing about the speeches and statements of both spacemen was how little specific information they supplied about their flights. Nikolayev, for instance, indicated he estimated the two spacecraft were about five kilometers apart at their closest point but preferred to say merely that he was delighted to see his friend with him in space "by my shoulder." This led to additional and largely inaccurate speculation on the flight achievements. As indicated earlier, their closest point was 6.5 kilometers (about four miles). Not mentioned at the press conference was the fact that Nikolayev had floated in his cabin, during four periods, for a total of 3.5 hours. Popovich had free-floated about three hours. Neither noted ill effects. The Russian concern about radiation later turned out not to have been justified, thanks, in part, to the way they programmed the orbits. Within a single day, the maximum radiation dosage for both cosmonauts was 11 millirad. Nikolayev, in his longer flight, received a total of 43 millirad, Popovich, 32 millirad. (The average person on earth receives about 29 millirad of cosmic radiation a year.) The Russians considered Nikolayev's 43-millirad absorption "absolutely safe."

When the parades died down, along with the world's journalists' periodic and largely superficial preoccupation with who was ahead in the space race, a more lasting fallout from the twin flight developed in the military sphere. The "space patrol," as one U.S. Air Force officer called the flight of *Vostok 3* and *4*, stimulated a reassessment of the military potential in space. Leaders of both the U.S. and the U.S.S.R. were already in the habit of telling their people, in almost identically misleading words, that their nation's goals in space were entirely "peaceful" or "scientific" when, in obvious fact, large and expensive program segments in both countries were substantially military. But the Nikolayev-Popovich near-rendezvous served to dramatize for the first time the enormous military potential of aiming one space probe at another body already in orbit. This line of speculation generated by the Nikolayev-Popovich space community again created for the Soviets a backlash effect that served to underscore certain U.S. Air Force budgetary priorities and that ultimately assisted in diverting more funds to American military space programs. So the Soviet "spectaculars" were already building a pattern of obvious paradox. Just as *Sputnik 1* rescued the U.S. from its "think small" attitude toward space, so did each subsequent Soviet achievement directly affect the American man-in-the-street, the military, and ultimately, the laws of Congress. The greater the Soviet propaganda achievement, the greater the response in the U.S. was to both "scientific" and "military" programs of its own. The more U.S. experts studied the dual flights, the more perplexing became the question of why the two spacecraft did not fully rendezvous. We had all the trajectory data. The void of space is the world's most transparent fish bowl. Distance lends high visibility. The orbital trackage circumnavigating the planet is as easy to plot and to project as an eclipse of the sun or moon. Deviations therefrom show up at Colorado Springs, at Jodrell Bank, at Bochum Observatory in Germany, and elsewhere as clearly as a Boeing 707 turning on its downwind leg on an overcast day at Kennedy International Airport. As hungry as the Soviets at first were for international acclaim, they may have reached the point, nevertheless, where they had to temper the acclaim in favor of the delayed issue of larger dividends. That point may have been silently reached, preceding and following the flights of *Vostok 3* and *4*. Soviet ground tracking, as evidenced by both pinpoint satellite landings and lunar-course corrections, was every bit as accurate as our own. They had more than enough weight available for maneuvering fuel. Orbital mechanics is incredibly complex, but its mysteries were certainly not incapable of solution by a mathematician, for instance, of the caliber of A. N. Nesmeyanov. Soviet scientific literature is replete with in-depth studies of the rendezvous required for assembling, staffing, and resupplying permanently manned space stations. The pioneering 1957 Soviet film, *With a Rocket to the Moon*, in which the best Soviet scientific consultants participated, clearly envisioned pinpoint rendezvous maneuvers as a prerequisite for the major, long-standing Soviet goals in space—exploration of the moon and planets. Yet neither *Vostok 3, 4, 5,* and *6* nor *Voskhod 1* and *2* exhibited the delicate maneuvers leading to rendezvous. And at the same time, Colorado Springs, augmented by our tracking facilities in Turkey and elsewhere, continued to shuffle to Washington classified reports that some Cosmos satellites—as far as we know, unmanned ones—were changing orbital plane and were altering their apogee and perigee and that a whole family of them—

known familiarly as the "eight-day satellites"—were being brought back to earth prior to the decay of their orbiting boosters. This mystery is a far more enigmatic one than that created by the more than two-year delay in manned flights that followed the flight of *Voskhod 2* on March 18, 1965. The Russians preferred never to aim to precede a U.S. accomplishment by more than a few months; there was still plenty of time to precede a U.S. manned lunar landing, but incredibly, the U.S. full rendezvous and docking had preceded the Soviet open demonstration of these vital techniques. The question of whether this precedence was deliberately granted or—based on Soviet faith in a lunar booster efficient enough for a direct flight to the moon—whether this step was deliberately bypassed is the great unknown of Soviet astronautics. Unless it was bypassed, its omission violates their otherwise unassailable chronology.

That is why most informed observers expected the next manned flight in the Soviet program—which, as Korolev said, never repeats a previous flight—to complete the process of docking and joining two vehicles in space by maneuvering one or both spacecraft until orbits and position coincide.

11
"I Am Seagull"

THE primary Soviet reason for sending a woman into space was not, as is commonly believed, propagandistic, although Premier Khrushchev readily agreed it was surefire internal and external propaganda.

The primary reason was biomedical. As gray-haired Academician Vasily Parin pointed out to me in Moscow with wry elementarianism: "Women are different from men. They are built differently. They have different organs. As you know, they menstruate. They sometimes have different reactions. We simply wanted to find out how this different half of the human race was affected by the space environment. We had experimented extensively by sending female animals, fish, and birds into space and now, we felt, it would be helpful to send a woman."

The decision to prepare women for an active role in space exploration was, partly, a natural consequence of the role of women in the U.S.S.R. It was, even more, a consequence of long-range Soviet space goals, which, from the beginning, have visualized women in lunar colonies, in permanent space stations, and as crew members of spaceships. As often pointed out, the women of the Soviet Union have held responsible positions and have adapted themselves to nearly every variety of human task, especially since nearly half of Russia's last half-century has called men to internal or external wars and to the battlefield grave.

In Soviet hospitals, there are more women doctors than men. More than 30,000 Soviet women are doctors or masters of science. Seventy per cent of all teachers are women, and 13,000 of these hold senior educational or administrative positions. The Supreme Soviet of the U.S.S.R. contains about 400 women, or about two out of five members.

When I visited the new space museum under construction at Tsiolkovsky's home town of Kaluga, women carpenters and masons in overalls and cotton shawls stared curiously from their workbenches at this alien American. At Novosti Press Agency in Moscow's Pushkin Square, I met a handsome 38-year-old mother of two who was the engineer in charge of Novosti's bank of Teletype machines. Near Novosibirsh at Siberia's "science city," I saw women running Mink-220 computers, and supervising the university and several research laboratories.

Since the U.S. at the time had no long-range space imperatives beyond "landing *men* on the moon within this decade," it had never occurred to any responsible NASA official that *Vive la différence!* might apply extraterrestrially. NASA, in fact, has always been very defensive about this subject, although astronauts say privately they would welcome women "with open arms." NASA's highly proprietary attitude about the domain of space has made many enemies in the scientific fraternity, but it was to find out its proprietary exclusion of bangles and bows and the unavoidable and stuffy implication that "this is man's work" eventually alienated American womanhood from Clare Booth Luce to Rosie O'Grady.

The female biomedical specimens Russia lined up in 1962 for the delicate selection process were all parachute jumpers. Some, but by no means all, had pilot training. Among them was a 25-year-old, slightly stocky and tomboyish girl named Valentina Vladimirovna Tereshkova. Valya, as she was called, had gray eyes, dark-blonde hair, and an engaging smile. At the time of the Gagarin flight, Valya happened to be walking by a row of looms at the Krasny Perekop Cotton Mill in the Yaroslavl region on the upper Volga. She was a graduate "cotton-spinning technologist," head of the Textile Mill Workers Parachute Club, and secretary of the Komsomol (Young Communist League) Committee. The announcement from a co-worker that Gagarin was in the Cosmos

A duplicate of *Sputnik 3*, the third Soviet space vehicle, weighing 2,925 pounds on display at the Science Pavilion of the Economic Achievements Exhibit in Moscow. Launched May 15, 1958, as an orbiting science laboratory, it confirmed the presence of the earth's outer radiation belt. Solar cells can be seen on side panels of the spacecraft.

A duplicate of the Vostok third stage and capsule that carried the first cosmonauts into earth orbit on display at the Leipzig Spring Fair in 1966. The thick metal sphere on the right, the cosmonaut capsule, detached itself prior to re-entry and landing.

A three-stage Vostok booster, 127 feet tall and more than 32 feet in diameter at its base, similar to those used in early manned space flights. The assembly contains 20 rocket nozzles for initial thrust. The first and second stages have a central component and four lateral components, each containing four nozzles, mounted in a cluster. The third stage consists of a propulsion unit, the space capsule, and a protective cone jettisoned just prior to the separation of the capsule from the propulsion unit.

brought tears to her eyes. The day's excitement was such that, she later confessed, she could not remember whether or not she had eaten dinner. When pictures appeared of Gagarin and his wife, Valentina Ivanovna, Valya recalls her mother commented: "A good couple. It's obvious they were born and bred among working folk."

At the time, Valya had made several dozen parachute jumps, which led her to imagine herself as a space pilot. In fact, a kidding girl friend called her "Gagarin with a skirt!" She was particularly delighted when Gagarin later told Queen Elizabeth of England that "there's no doubt a woman will fly in space." She dreamed of herself more and more as a space flyer. Valya literally threw herself into perfecting her jumping technique. Once in a tournament her instructor told her she landed "like a bear." She went off by herself in the weeds and cried as she plucked out daisy petals one by one. But on her next jump she earned a "fine" by hitting 65 feet from the landing cross. After the Titov flight, the young textile worker impulsively wrote a letter to Moscow indicating her desire to train for space flight. It turned out to have been a happy impulse. The following March she was asked to report to Moscow, where she and other women cosmonaut candidates were flown to "Star Town" at Tyuratam. Her determination and qualification were already known there, as the two cosmonauts who escorted her to her room were the next two to fly in space, Pavel Popovich and her future husband, Andrian Nikolayev. As Nikolayev set down her suitcase in her room, she recalls "the sparkle in his dark, velvety eyes."

As training of women cosmonauts proceeded to the centrifuge, the running belt, and other tests, she discovered a difference of opinion about how they would function in space. "Nobody knew how a woman's organism would react," she said. "The doctors and biologists were divided between the two viewpoints. Some maintained that the woman's organism would adjust to any situation. Others adhered to the opposite view and claimed that the 'weaker sex' was less adaptable to the varying conditions of a space flight."

"Cold did not trouble me," she said of her training in the

thermochamber. "In the village where I was born I used to spend all my winters out in the fresh air dressed lightly, even in icy wind. But for a northerner like me, heat was harder to stand . . . as the sweat poured down in torrents, I dreamed of a choc-ice or a drink of some cold village *Kvass*."

Valya especially feared the soundless solitude. She did not doubt her courage and reminded herself how many times as a child she had run across her village cemetery at night on a bet, but she had feared that a woman was less stable mentally than a man and that solitude was more difficult for a female. During her solitary and soundless confinement, she longed for "the buzz of a fly or a mouse scratch somewhere." She managed to survive her worst periods by reciting Nekrasov's poetry. Nekrasov had often celebrated in his poems the emancipation of Soviet women since the Csarist days of which he once wrote: "and each of these burdens has heavily lain on women of Russia's domain."

It is not known which of Nekrasov's poems were her favorites, but she certainly had read the following one. It seems to fit the tomboy from Yaroslave:

> In out-of-the-way little corners
> Those strong stately women are seen
> With dignified calm on their faces,
> The gait and glance of a queen . . .
> At games not a youth can outstrip her
> Misfortune her nerve will not tame,
> A run-away horse she will master,
> Walk straight to a hut that's aflame.

Valya took eagerly to advanced parachute training, working under the same expert instructor, Nikolai Konstantinovich, who had trained Yuri Gagarin. On her 100th jump, she landed in an abyss, but Konstantinovich sent her back for a series of 26 other jumps, each more difficult and demanding than the preceding ones. Contrary to a number of reports in the U.S. that Valentina Tereshkova flew in space without first learning to fly a plane, she received rather extensive pilot training after entering the cosmonaut training program.

She flew first as a co-pilot on a transport, then progressed to the piloting of jets, which she flew at times over her Yaroslave region.

Like her male predecessors in space, Valya also met chief designer Sergei Korolev. "All during the ride to the plant," she recalled, "I kept wondering how the chief designer would receive us girls. He turned out to be an unassuming man with a keen sense of humor and very sociable. He knew all about us, our names, where we were from, our education. He also talked about the progress we were making in training. He took one entire group through the airy shops of the plant. He introduced the engineers and workers and showed us a spaceship . . . all during our interview energetic young people kept coming up to him with blueprints and questions. . . . We would like to have asked him hundreds of questions and to spend the entire day with him, but he took a side glance at his watch. It was time for us to leave."

Later, when Valya began Vostok-simulation training, her rigorous schedule helped trim down her figure, which normally gave her the appearance of a slightly chunky Ingrid Bergman. In early June of 1963, a select group of men and women cosmonauts flew to Moscow for the traditional and secret preflight meeting with the Chairman of the State Commission and key members of the Communist Party. The meeting was already under way when Valya entered the rear of the room unnoticed and took her seat near a window. Later, the chairman looked directly at the delicate-appearing girl and announced: "I name you, Valentina Vladimirovna Tereshkova, to command the spaceship, *Vostok 6*." A writer present later described her appearance and reaction to her selection: "When she stood up, she was immediately liked by everyone. She had a spiritual face with delicate features. She had luxuriant hair which lay in soft curls. Her eyes peered out from underneath narrow eyebrows. Their expression was soft and somewhat confused. But then she spoke.

" 'I am proud that I, a simple Soviet girl, have been given such trust.' Her voice was soft and calm. It did not show excitement. Only her face blushed."

Also officially named at the meeting to be pilot of *Vostok 5* was a 28-year-old jet pilot and former parachute instructor, Valery Fyodorovich Bykovsky, who had been backup pilot to Nikolayev. By coincidence, his wife, a student of history, was also named Valentina. They then had one son, Valery. Valery, the newly announced space pilot, was closer in temperament to Gherman Titov than to the other cosmonauts. As a youth, he broke windows with a football and once, during training, he had complained that his instructor was bogged down in red tape and was a "slave" to military discipline. "One day you'll be ashamed of your exasperating behavior!" his instructor was reported to have replied. Titov, who had a red-haired temperament himself, once described Bykovsky as "like a boy, very hot and curbs his desires only with the greatest difficulty." Friends reported that he was so moved by the birth of his son that he wept.

A near-Muscovite, Bykovsky was born in the small suburb of Pavlov-Posad. He had been the first cosmonaut to try out diabolical Soviet space-training devices and held the Order of the Red Star prior to his space flight. He was the first cosmonaut who had not joined the Communist Party. He is broad-faced, has a wide, serious mouth, and a cowlick.

After the "Romanov and Juliet" team of space pilots received the congratulations of fellow cosmonauts, scientists, and dignitaries at the secret meeting, they went anonymously to the highly public Lenin Mausoleum in Moscow's Red Square. With them were the secretary of the Komsomol Committee and Valentina's backup cosmonette, who has never been identified. As they stood in silence for the traditional farewell, they were not recognized.

The next morning a plane took them and the other cosmonauts to Tyuratam. En route, Valya, her face against the window, was impressed by the colors in the clouds below.

"How beautiful it is all about!" she exclaimed.

"Wait a bit," echoed her future husband, Andrian Nikolayev, "you haven't seen anything yet."

She reported later that Nikolayev was particularly attentive to her on that last airplane flight. "He bent every effort," she said, "and advised me to rest at every opportunity and not to lose my head. And I knew that this peace of mind was his second wings, and I attempted to imitate him to some degree."

When they deplaned at "Zvesgograd," they found a still-expanding and fast-paced "city of stars." Everywhere there were changes. New houses were marching out across the windy steppes. Entire housing developments as large as Moscow's Cheremushko were under way. Zvesgograd now had a brick music school, a technical school, a new "Berezka" café, and a large nursery on the outskirts. The advertised movie at one of the two theaters was *A Concoction for Marriage*. The main hotel was teeming with people who were assembling for the imminent launch. That night the cosmonauts danced under the moonlight to waltzes from a portable radio played in the garden. Nearby, the bushes the cosmonauts had planted on the eve of Gagarin's flight now had berries on them. As Valya danced, Bykovsky showed friends the color photograph of his son he planned to take with him into space, along with twelve Komsomol buttons. Before the gay evening was over, friends talked him into also trying to smuggle aboard some berries from the two-year-old bushes.

The next day and night the warm, spring desert wind was high, blowing sand across the launch pad where Bykovsky's poised *Vostok 5* waited. Because of the winds, Korolev postponed the flight, which freed all the cosmonauts for their traditional prelaunch fishing party on a nearby stream bank. That afternoon and evening, Gagarin took charge. As "chief design engineer for the ground situation," he divided up duties among the cosmonauts. He demoted the newly assigned space pilots by designating Valya and her substitute cosmonette to peel potatoes and appointing Bykovsky to be cook. Nikolayev was made senior fisherman. In the king-of-the-mountain horseplay and confusion on the riverbank, the laughing cosmonauts threw an unknown young man in the water with all his clothes on. To the embarrassment of all, he turned out to have a soaked document in his pocket designating him as a "doctor of technical sciences," a prestigious title in the Soviet Union. The party caught an abundance of fish, which Nikolayev converted into a stew in a kettle hung from a tripod.

On the morning of June 14, the wind had died. There were no clouds and the sun was scorching the huge concrete slab at the base of the *Vostok 5* launch pad. Korolev, wearing a straw hat, dark glasses, and a white silk shirt with its tail out, gave quiet orders that were carried out instantly. "Sergei Petrovich," he said at one point, "the work schedule must be stuck to." "It will be done," came back the reply. Shortly after a hot noon, Bykovsky's bus, filled with singing cosmonauts, deposited him at the rocket's base. He had slept later than anyone else that morning—until 8 A.M. After his formal and traditional speech, he entered the elevator. "Until my return to earth," he saluted. Then he rode to the top of the rocket. Once inside, he conversed by phone with Korolev and Gagarin as distinguished visitors watched from the usual post, the old-fashioned, second-story veranda several hundred yards away. They heard the periodically announced time to launch, as at Cape Kennedy, until liftoff at 3 P.M.

An *Izvestia* reporter, G. Ostroumov, later reported the sensation of launch: "I am writing these lines under the thunder of a rocket. A spoon is shaking in a glass which stands near me. Above my head the roof of the veranda is shaking . . . I see the inversion trails turn white as if they were drawn out of the heavens . . . and then the shining dot disappears, lost in the emptiness of the blue. And the roar still comes from the heights. And the team draws the ship with twenty million horsepower."

Watching from another observation post was Valentina Tereshkova, who unconsciously scratched a groove in a wooden barrier with her fingernail at the instant of liftoff.

Bykovsky's first words under his code sign of *Hawk* were "Everything is fine," as his pulse shot up to 106 beats per minute, four beats lower than John Glenn's had been. Shortly thereafter, he was in a secure orbit with virtually the same inclination, apogee, and perigee as those of *Vostok 3* and *4*. On the ground, a calm and expectant Valya, who had had her hair done earlier that day, took the microphone: "*Hawk,*

Hawk. Do you recognize my voice? A warm greeting for you . . . I congratulate you on a good beginning."

Valery Bykovsky smiled at her from the television screen. "I am waiting," he said.

There was plenty of time. Those watching her appeared from their reactions to be contemplating a Joan of Lorraine. "Tereshkova is next to the rocket," one man wrote. "A severe giant rocket and she is dressed in a light blue dress, and white shoes, a girl—the elegantly dressed young girl who will take the rocket. . . . Women had said good-bye to their men before going on a flight and they had waved their hands in farewell. The men flew off but the women remained on the ground and waited. And now the men would say farewell to a woman. With tenderness and respect they give her flowers. Fly, *Chaika!"*

A middle-aged scientist remarked: "How she behaves! A woman but in terms of her self assuredness and quiet is not inferior to men. True, not inferior."

The next day, June 15, a new life and new tempo began to build at the Soviet space center as *Vostok 6* was readied. During the day, Valya and her backup moved to the brown cosmonaut cottage. An old, gray-haired woman greeted them. "I am happy to see you, my girls," she said. Then she pointed to a sturdy iron bed. "Take Gagarin's bed, Valyusha [Valentina]. You are also the first." Valya noticed that the old woman, who had also looked after each preceding cosmonaut, had placed her favorite flowers, white gladiolus, on the table.

The next morning Valya put on lipstick and face powder as usual, and picked up a small red flag her mother had woven for her to take. Then she told the old woman good-by. Later, after her physical exam, she carefully dressed in an orange space suit that carried something new, a snow-white dove embroidered over her left breast. A man who watched her walk toward her rocket noted quaintly that "her space clothing stressed her strict simplicity but at the same time did not hide her tenderness." She kissed Gagarin, Titov, and finally, her husband-to-be, Andrian Nikolayev. Later when she took her seat in *Vostok,* poised on the same pad that had launched Bykovsky, she reported immediately: "I am *Chaika* [*Seagull*]. I feel fine." Korolev responded: "I can only envy you. We feel worse. The sun is scorching, and why should we stay on the earth." The outside temperature at the desert base was 95°.

An observer, N. Melnikov, later noted: "Valya waited the launch in an elevated mood, as if it were easy. One could feel that she grasped everything for the flight and had a good mastery of her equipment. . . . The earthly expectation lay heavily on us, and this strain grew. And she launched peacefully [at 12:30 P.M.]. She noted, almost to herself, 'I'm off.' And all of a sudden her voice changed, and rose in excitement: 'I see the horizon. A blue blue strip. This is the earth. How beautiful it is.' "

Her "space brother," as Russians were now calling Bykovsky, was now coming overhead on his 31st revolution. In the Tyuratam underground control center, his face came on the television screen. Instruments now rectified Valya's launch time to the one-thousandth of a second as plotters traced her rising trajectory toward the approaching orbit of *Vostok 5*, coming up from the equator. It is not known what Valya's pulse rate was during her climb out, but it soon settled down to a near normal of 80 per minute, with a respiration rate of 20 per minute. On her first orbit, *Vostok 6* reached its closest point to *Vostok 5*, 3.1 miles away. Valya's apogee and perigee were just about a thousand yards outside those of Bykovsky. Small manual maneuvers were carried out, but—probably due to her slightly slower speed—the distance between them gradually increased rather than diminished as both orbits decayed.

As soon as they were together, they sent a joint message: "Have started carrying out joint space flight. Dependable radio communications established between our ships. Are at close distance from each other. All systems in the ships are working excellently. Feeling well." On her fourth orbit Valya talked to Khrushchev, and Russian and East European television viewers cheered when she smiled and waved from space. As the dual flights continued, the now-bearded Bykovsky was seen briefly as he dined on roast veal, white bread, cheese paste, and black currant juice (he later ate caviar and sausage), but the drama and interest were focused on the dimpled maiden from Yaroslave. Even before she landed, groups in Russia marched with placards reading "Glory to the Explorers of the Universe" and shouted in unison: "Cosmos! Cosmos! Cosmos!"

Valentina Tereshkova, who overslept her first space nap, continued to be the center of mounting interest even as Bykovsky reached and exceeded Nikolayev's record of 64 revolutions. On June 18, when she passed Gordon Cooper's American record of 22 revolutions, it became clear that her flight had gripped world imagination even more strongly than the 1890 round-the-world flight of Nellie Bly. In response to concerned queries, Academician N. V. Parin stated that her present experiences would not impair "the principal functions of the woman's organism connected with child bearing," and that "woman's organism is in no way second to man's in its physique. Her specific physiological features do not limit her ability to fly in space." Soviet physiologist Pavel Vasilyev had been concerned on previous flights that blood pressure sometimes fell and that the muscular tone of the heart had become weaker. Interest in Valentina Tereshkova's personality and daring almost eclipsed the fact that both spaceships were crammed with scientific instruments and experiments—especially vestibular and physiological ones.

Somewhat earlier than Soviet officials in Moscow had predicted, a BBC monitor in England picked up a Tereshkova message: "This is *Chaika*. I am ready to record data on manual descent." On her 49th orbit, she fired retros, and landed by personal parachute in a flat field 385 miles northeast of Karaganda, where *Vostok 3* and *4* landed. The only injury was a slightly bruised nose. Two hours and 46 minutes later, Valery Bykovsky, in his fifth day and on his record-breaking 82nd orbit, landed—also by personal parachute—335 miles northwest of Karaganda. He was in excellent condition. Amid some feeling of letdown that a full rendezvous had not been achieved and that Bykovsky had not gone the full predicted eight days, the Russians simply went wild over

F*

their instant Joan of Arc. In the final analysis, it was the unassailable biomedical logic of her flight and the surprise of it coupled with the astounding propaganda spin-off from the first space journey of a woman that provided the Russians with such a transcendent achievement. The worldwide reaction to Valentina Tereshkova's flight was far more favorable than even Khrushchev could have anticipated. The Swiss newspaper *Die Tat*, acknowledging that the Russians have a very special sense of mass psychology, correctly assessed that "they hit the apathetic nervous system of people today." Queen Elizabeth of England sent congratulations. Women Olympic champions the world over rallied to endorse Tereshkova's feat. "I admire that Russian girl. I think she is great," said Fanny Blankers-Koen of Amsterdam, an Olympic medal winner. "To the millions of underprivileged women of the world," wrote the London *Daily Express*, "she is a soaring symbol of feminine emancipation."

Russia had so timed the flight that it preceded by one week a worldwide conference of women scheduled for Moscow. On Ladies' Day in Moscow, the ladies made the most of it. Those who attended the postflight private and public celebrations said they exceeded in enthusiasm, in national fulfillment, and in sheer bedlam any previous space-flight celebration.

In the U.S., NASA attempted to keep the same tight lip it had always maintained on the subject. NASA had no immediate or future plans for women space pilots. However, two American women who had tried long and hard to become astronauts chose the occasion to speak out against the "For Men Only" policy. Pilot Jeraldine Cobb, 32, of Ponca City, Oklahoma, had been accompanied by a *Life* magazine correspondent for months in the expectation that—somehow—NASA would let her fly. She said if NASA would just make up its mind, it could "launch a woman within months. After three years I had more or less become resigned to the fact that since we were doing nothing in this area, Russia would . . . I'm proud that a female made the trip but I'm disappointed in the nationality." Another woman who had passed astronaut physical examinations at the Lovelace Clinic in

Albuquerque, New Mexico, along with nineteen other women pilots, was Mrs. Philip Hart, wife of Senator Hart of Michigan. Said Mrs. Hart: "I'm more annoyed at the fact than I am impressed. It only showed that the U.S. is 100 years behind in using the full abilities of women."

Russia also rode the exultant wave created by the romance between Valentina Tereshkova and "the most eligible Russian bachelor," Andrian Nikolayev. Citizens eagerly followed their November marriage in the first official public wedding ever held in the Soviet Union. Celebrations broke out again when their healthy daughter, Yelena Andrianovna Nikolayev, was born the following June. Khrushchev had told them at the wedding, after drinking twenty-one toasts to them, that "if you have a baby, the fifths won't fail to come."

Khrushchev made another prediction: that the pair would "have other flights." Later, in the spring of 1967, a widespread and often-repeated rumor in normally responsible quarters suggested that Russia has given up plans for a manned flight to the moon. But Valentina Tereshkova did not share this view. "The moon team has already been picked," she said on her trip to Havana. "Major Yuri Gagarin is head of it, and I am on it."

12
Flights of the Voskhods

WHEN the manned, single-seat Vostok series and the American single-seat Project Mercury series ended in the spring of 1963, Russia had launched six spacemen into orbit to our four (Glenn, Carpenter, Schirra, and Cooper—the Soviets do not regard the suborbital shots of Shepard and Grissom as true space flights). The Russians had done every essential thing first and held a fistful of records that indicated, both internally and externally, that they had gained rather than lost ground in manned space-flight experience—although the U.S. was fast catching up in unmanned, brilliantly instrumented flights. Soviet enthusiasm for space was at an all-time high. After the Tereshkova flight, for instance, one exuberant Russian citizen recognized an American correspondent during one of the parades in Moscow. "Why are you trailing us in space?" he taunted. "Why do you not catch up with us?" In this atmosphere, Khrushchev had little difficulty persuading either Soviet officials or the people to make additional sacrifices for the conquest of space.

The manned program of both countries concentrated thereafter on preparations for orbiting larger spaceships with multiple crews. For America, this meant nearly a two-year pause in order to prepare its new, 7,000-pound, two-man Gemini (Mercury weighed 3,000 pounds) for flight experience. Before we could orbit our first astronauts (Virgil Grissom and John Young) in Gemini, however, Russia put Sergei Korolev's new, multiseat Voskhod (meaning "Ascent") into two new flights that continued her unbroken precedence of manned space events.

The *Voskhod 1* orbital payload, designed by Sergei Korolev, called for a weight of 11,730 pounds, and to loft it, the Soviets needed a booster stronger than the estimated 1,340,000-pound-thrust booster that sent the Vostoks aloft. It had taken the Russians twenty-seven months to revamp and man-rate the large, relatively inefficient, but reliable first-generation ICBM for the Vostoks. For the Voskhods, they needed and got a new clean-lined, silver-colored, liquid-propelled ICBM booster that had an estimated thrust of 1,430,000 pounds. It was undoubtedly revamped and man-rated by the addition of the most powerful second stage then in the Soviet Union.

The Voskhod that Korolev and his colleagues produced was a vastly improved, flexible, commodious, and comfortable spaceship. Like Vostok, it was shrouded in a pointed air shield for its ascent through the atmosphere. When the streamlining was jettisoned, what was left was a rounded spaceship from which antennas and the parachute canister jutted at the top and the retro-rockets jutted at the bottom. Unlike Mercury, Gemini, or Vostok, it had no ejection system for its crew. It was designed to land with its encapsulated pilots on land or water by a combination of parachutes and a second, or reserve, retro-rocket fired just before impact. When I met and exchanged space souvenirs with cosmonaut Colonel Vladimir Komarov, the commander of *Voskhod 1*, he demonstrated how the rocket brake permitted a gentler landing than the braking of a modern elevator. The unsinkable and hermetically sealed craft and its relatively low-oxygen gas system also permitted a shirtsleeve environment for crew members. It was possible, for instance, to smoke in Voskhod, had crew members wished, and space suits were not necessary. (Strictly speaking, space suits were not necessary in all predecessor spaceships either; they had been worn, however, for safety.) Its air-regenerative system was basically an enlargement of the Vostok chemical system. Air pressure was normal, temperature fluctuated between 18° and 21° centigrade and humidity between 45 and 60 per cent.

Voskhod had a new system of orbital-flight orientation that cosmonaut-scientist Konstantin Feoktistov called obscurely "ionic velocity vector formers." Voskhod was also a better-insulated spacecraft. Its relatively uncomplicated interior was padded with snow-white porolon, which helped reduce the noise of liftoff to below the level inside a jet on takeoff. Voskhod's instruments were more numerous than Vostok's (many were the same) but far fewer than those of Mercury or Gemini. On the center panel, as in Vostok, were the globe of the earth, the ship's clock, and basic flight instruments. Next to the commander's contoured seat on the port side was the black plastic control handle. A conveniently located oblong panel contained tumbler switches, pushbuttons for various systems, a radio station, and a telegraph key. Nearby was a locker for water and space food and another container for warm clothing and life jackets, in case the spaceship landed on water. On the glass porthole beside the starboard seat was mounted a vision device for studying the earth and stars. The three seats were arranged in a horizontal row, and each crew member had his own viewport.

Six men were trained specifically for the flight of *Voskhod 1*, although their training skipped certain things, including the chamber of horrors, and was not as intensive physically as earlier cosmonaut training. Two of the men were jet pilots, two were scientists, and two were doctors. One-half of each pair formed the backup crew.

Picked as spaceship commander of the prime "new troika" was a 37-year-old Red Air Force pilot, Vladimir Mikhailovich Komarov. When I talked with him in Moscow in June of 1966, neither of us knew, of course, that he was to become the first cosmonaut to make two space flights and that his second one was to end tragically. He was a slender, athletic man with dark brown eyes that lighted up when he talked and very white teeth that flashed when he smiled. He was born in Moscow and, as a young man, worshiped such famed Soviet pilots as Chkalov, Gromov (who flew to the U.S. over the North Pole), and combat pilot Victor Talalikhin, who rammed a German *Heinkel* bomber. He entered pilot training at fifteen in World War II but graduated too late to see action. He continued training at the Third Sassov Air Force School, at the Chkalov School in Borisoglebsk, the Serov Flying School in Bataisk, and the Zhukovsky Air Force Engineering Academy in Moscow, which corresponds to the U.S.A.F. Institute of Technology at Wright-Patterson Air Force Base, Ohio.

Before his space flight but during his training, Komarov was grounded for a systolic heart murmur somewhat similar to the condition that permanently grounded U.S. astronaut Donald Slayton. Yuri Gagarin knew Komarov then and tried to encourage him. "Volodya decided to prove he was in good health," explained Gagarin, "and that he had a right to occupy a seat in the spaceship cockpit. His organism was carefully examined in the hospital. Everything was found to be all right, but the doctors could not forget about the systole and they did not say 'yes.' Volodya took leave and traveled to Leningrad to see the famous specialist, Professor Molchanov. The specialist agreed there was no danger in the irregularity." Komarov continued to go to the highest specialists in the Soviet Union and was, finally, thanks to the medical endorsements he accumulated, readmitted to the space-training program.

His wife, Valentina, mother of their two children, worked as a librarian in the cosmonauts' village near Moscow. Here cosmonauts relaxed with games that Komarov told me about.

At the time I interviewed him in Russia, I had just lost an assignment from *Life* when the U.S. State Department, at the last minute, refused to let my friends and neighbors, astronaut Scott Carpenter and his lovely wife, Rene, continue their plans to visit the Soviet Union. I told Komarov I was to have joined the Carpenters at the Moscow airport and report on their reaction to the Soviet Union, to the cosmonaut village, and vice versa. Komarov was obviously keenly disappointed. He explained he had had a particular affinity for Scott Carpenter ever since he had read of his motorbike accident in Bermuda, which had permanently grounded him from space flights. "I sympathize with the way he feels on two accounts," Komarov said. "First, because I was grounded for this heart thing and, second, because I broke my nose in an accident in cosmonaut town. They gave us a choice of three group physical activities. Ballet, which we turned down quickly, soccer, or Canadian ice hockey. When we chose Canadian ice hockey, they warned us to be careful. At our first real game with cosmonauts playing on both teams, they drove up two ambulances and the doctors lectured us again on safety. At first we started out very slow and gentle as they had advised us. Then somebody tried hard to make a score. Suddenly, bodies were thumping and sticks were flying all over the place. And I got my nose broken and thought it was my head and that I would be grounded again. Tell Commander Carpenter I know just how he feels. If he, or any other cosmonaut, ever gets to Russia, we'll try to show him a good time. We greatly respect the American cosmonauts."

One of Komarov's extracurricular activities was his editing role at Novosti Press Agency, where he contributed to and supervised publication of articles and books on space activities, many of which were useful in the preparation of this book.

Also picked for *Voskhod 1* was the first scientist-cosmonaut, Konstantin Petrovich Feoktistov, 38, a blond, long-haired, serious-minded intellectual who is very fond of chess. At sixteen, Feoktistov volunteered for army service when the Germans threatened his home town of Voronezh, where Peter the Great built the fleet used against the Turks. While on a reconnaissance patrol, he was wounded and nearly killed by a Nazi SS officer. During his convalescence, Feoktistov entered one of Russia's best engineering schools, the Bauman Higher Technical College in Moscow. He earned his candidate's degree (Ph.D. equivalent) six years later. He worked as an industrial engineer, then joined the space program as a specialist-lecturer to the cosmonaut class. As a calm, knowledgeable lecturer, he became a cosmonaut favorite. They were delighted when their teacher moved to their side of the podium to train himself for the first flight of *Voskhod*.

The third assigned crew member was a 27-year-old, short-haired doctor who somewhat resembles the actor Cornel Wilde. Boris Borisovich Yegorov, the son of a prominent Soviet brain surgeon, was one of the specialists in aviation medicine who met with Yuri Gagarin following his flight. When he was eighteen, he entered the First Moscow Medical Institute to become an aviation medicine specialist, with a particular interest in the vestibular apparatus of the human ear. He became fascinated by complex medical equipment and invented a radiosonde for examining the stomach. During space-flight training, he made eleven parachute jumps. He was somewhat used to heights as an experienced mountain climber. Before his flight, he had made thirty ascents in the Caucasus Mountains near the Caspian Sea. His pretty wife, Eleonora Victorovna, mother of their one son, was a doctor in a leading Moscow eye clinic.

Just before Columbus Day in October of 1964, the three men moved into the now-cramped quarters in the traditional brown cottage, or cosmonaut bunkhouse, at Tyuratam. Only Komarov had slept there before, when he had been backup for Pavel Popovich. On the eve of his own flight, he chose his former bunk. The iron bed of Gagarin and Tereshkova went to Feoktistov, while the lanky Yegorov had the small adjoining room to himself. Other cosmonauts and friends soon joined them and threatened to stay all night, until a serious, bareheaded figure in a light topcoat appeared at the door. When they recognized Sergei Korolev, they all stood up quietly.

"What a din," he said, assuming the uncharacteristic role of taskmaster, "they can hear you all over the cosmodrome." The noisy guests departed, and shortly thereafter the three crewmen stepped outside to look at the stars. The night was clear and cool. This was the latest in the fall that the Russians had ever attempted to put men in orbit, and the snow in the Urals was already lowering to the flanks spreading to the endless steppes.

Monday, October 12, dawned for the troika more like a holiday Sunday. As they exercised on a nearby hard-topped road lined with neat rows of poplar trees with whitewashed trunks, a single guard watched them from a barricade 100 yards down the road. After showers and breakfast, they dressed in just ten minutes in lightweight wool trousers and shirts and white, lightweight earphone helmets. They put on light-blue jackets before they entered the bus.

Komarov later said, "there was an awkward moment on the bus to the launch pad. Somebody suggested, as usual, 'let's sing!' We made hopeless gestures. None of us could sing. We were sorry not to have had Pavel Popovich with us. He would have given us a lead at once. But as it was, we could do nothing but joke about it." To conserve flight time for their full schedule of duties, they ate their lunch in the bus from red tubes.

The cosmonaut launch pad at Tyuratam consists of a truly massive, four-pronged service structure that, nevertheless, reaches some twenty feet short of the pointed tip of the rocket. The elevator stops short of the top, and there is a short flight of steps, known as "the cosmonauts' walk," to the top. The imposing gantry appears to lean against the rocket like a huge flying buttress, and the elevator, which first carried up the two technical-crew members, was smoothly winched upward at an angle of about 50°. Finally, the commander, Komarov, stepped out of the cold wind and ascended in the elevator. At the top, they removed their blue, zippered jackets and boots and donned light, suede slippers. Yegorov entered *Voskhod* first, then Feoktistov, then the commander. Without space suits, they looked, as they strapped themselves in, like three pilots ready for a session in a Link trainer. Korolev and a young engineer known familiarly as "the shipbuilder" paid them their traditional last-minute visit. When the hatch was closed, *Voskhod* became warmer. From his post as capsule communicator, Yuri Gagarin gave them ground progress reports through a Voskhod loud-speaker.

What they later described as "the incomparably grandiose spectacle" of blastoff came at 10:30 A.M. They all glanced at each other; then, as the rocket moved under them—perhaps reflecting the security of a group—their pulses rose only slightly above normal, noticeably lower than that on previous solitary flights. Within a few minutes, they were in orbit, but not precisely the one aimed for. Their new booster and guidance did not yet have the accuracy of the reliable *Vostok* booster. The separation of apogee and perigee, instead of on the order of about 30 miles, was 55 miles. The low point of orbit, around 100 miles, dragged more than 10 miles deeper into the atmosphere than the Vostoks had. This created friction, which—while not enough to jeopardize initial orbits—nevertheless caused more drag than the Soviets had planned.

There was no immediate danger, and Alpinist Yegorov studied the peaks traversing far below them. After the initial orbit, each time *Voskhod 1* passed over the green hills of Africa he looked for Mt. Kilimanjaro, of which he had read in the work of his favorite author, Ernest Hemingway. One evening they interrupted a meal of fried chicken and vegetables to study night clouds far below, which pulsed with the downy orange glow of a violent internal thunderstorm. They were far from the range of its deadly bolts or the sound of their crashing fury. They were within range, however, of the bullets of space—meteorites. And these they saw far below them, flashing in from all directions, as they burned in a chaotic pattern of white streaks within the unseen edge of the atmosphere.

Throughout the flight, Yegorov was the busiest of all, as he sought to take maximum advantage of the presence of a physician in space. Soviet investigation of the body in space has been more thorough and detailed than our own, primarily because our controlling medical objectives during manned flights were shortsightedly limited largely to verifying that a man could survive a flight to the moon and back. The Soviets were more scientific about the subject primarily because they were interested in physiological and psychological effects of far more subtlety than we were because their sights were, from the beginning, on far more ambitious roles for men (and women) in space. The cheerful—and to the Soviets, misleading—optimism of astronaut physician Dr. Charles Berry at televised Gemini press conferences reflected our unfortunate and abbreviated view of the function of space medicine. The smiling and personable Dr. Berry made a hit with television and other audiences when he said again and again that he foresaw no medical problem for man in space. But he did a disservice to the cause of medical science in his implied suggestion that limited investigation could produce an unlimited prognosis. It was clear that Dr. Berry's view of man in space was limited to the brief round trip to the moon with which he was chiefly concerned and for which NASA had been prominently and dramatically budgeted. Soviet Doctors Parin, Gazenko, Gorbov, Karpov, Yemelyanov, Saksonov, and their colleagues suffered from no such myopic view. While we looked symbolically toward a satellite, they were looking toward the planets. From the beginning, they investigated minute aspects of the human organism with the same thoroughness with which they had pioneered the medical investigation of animals in space. They appointed and regarded Dr. Boris Yegorov chiefly as a portable medical investigator to supplement *Voskhod*'s extensive biotelemetric systems. They had him carry aboard a complete investigative kit, including removable transducers and electrodes. And they gave him plenty to do. He experimented with noise devices, testing acoustics at a wide range of frequencies. In addition, he worked with electrocardiograms, seismocardiograms (to measure fluctuations of the chest cavity caused by heart action), pneumograms, electroencephalograms, and electroculograms. He made observations on pulmonary ventilation, conducted investigations of the functional state of a vestibular analyzer, and supervised and performed experiments involving the strength of the hand in rhythmic depression of the spring in a small dynamometer. All experiments were valuable, but the latter one established the surprising fact that the strength of the hand is increased in the weightless state.

The seemingly simple dynamometer test was particularly valuable, since its performance was measured by the aforementioned medical analyzers, which simultaneously monitored all major functions of the body. The test not only showed the precise point at which fatigue set in, but was also able to distinguish between mental and muscular fatigue.

Yegorov was the only man aboard to test himself in the motor act of writing such simple symbols as double spirals and the figure six. By performing the test with his eyes both open and closed, he established significant comparisons between ground and space environment.

Yegorov also had with him a portable adaptoreservometer for measuring the light sensitivity of the eyes and a Herschel prism to evaluate the tone of the eye muscles. Numerous blood tests taken in flight were later analyzed for the content of chlorides, sugar, urea, cholesterol, proteins, red and white corpuscles, and the disturbances of walter-salt exchange during weightlessness. He paid particular attention to his specialty, the study of the vestibulary apparatus, whose disturbance can cause dizziness, nausea, sweating, and vomiting. Periodically, crew members were required to rotate their heads ten times while hooked up to a battery of medical investigators that were accurate enough to distinguish between sensory and somatic reactions and the more gradual disturbance brought on by repeated stimulation. As both subject and doctor, Yegorov was also able to experience and analyze the spatial disorientation that Titov had complained of. Beginning with his second orbit, Yegorov felt a warmness in his head as blood accumulated there and, shortly thereafter, he had the uncomfortable feeling that he was flying through space upside down. Feoktistov was somewhat similarly affected, although the duties of neither were severely impaired.

In fact, when their first twenty-hour hours in space drew to a close, all three of them asked to stay up longer. "Feel fine, wish to extend observations, seeing much of interest and beauty," *Voskhod* radioed to Sergei Korolev.

Korolev replied, "There are more things in heaven and earth, Horatio. . . ." They correctly translated the chief designer's answer to mean no. "We began our preparations with a certain amount of ill grace," they recalled later. Preparing Voskhod for landing was something like preparing Mercury or Gemini. Instruments, especially medical instruments, had to be stowed, and the cabin had to be secured for three impacts: from the atmosphere and sonic barrier; from the jerk of the parachute; and finally, from the earth itself. They took this occasion also to autograph some patriotic and other souvenirs. They signed an original composite handwriting sample of both Lenin and Marx "9:15 A.M. Oct. 13, 1964, on board the spaceship, *Voskhod*."

On their 16th and final orbit, they tightened their straps over the huge island of Madagascar. After passing over Africa, on course, Komarov assumed re-entry attitude and the first retro-rocket fired. With a clang, the retro device separated, then continued to fly alongside as *Voskhod* slowly rotated. Minutes later, they saw their ports turn various shades of yellow, then finally assume the color of a carrot as their heat shield glowed under 10,000° F of temperature. Only inches from this scorching halo of fire, they were quite comfortable. As G forces mounted, "We all realized what was happening. There is a slight vibration, as if a good car driving at high speed has suddenly passed from an asphalt to a cobblestone surface. *Voskhod* is passing back through the sound barrier; its speed has become subsonic." Next they heard a second, dull metallic clang, as the parachute hatch flew off on schedule at an altitude of about three miles. Then the parachute jerked them back to terrestrial gravity, as they continued their descent at a speed of 220 meters per second. A few minutes before they had looked down on Africa steaming in the sun; now they drifted to earth through clouds that were spewing snow. The retro-rocket, triggered by its proximity to the ground, fired, and they landed softly at almost zero speed. They heard dry rubble crackling outside the spaceship's skin. Komarov released the parachute, so *Voskhod* would not be turned over or towed. Then he opened the hatch, and they felt a sudden surge of cold wind and blown snowflakes. As they stepped out to embrace and dance around, the three men already heard the helicopters that were homing in to their landing point. They came to earth not far from the Urals in Kazakhstan, about 184 miles northwest of Kustanay.

Following the successful flight, the usual round of rumors appeared in print. The most persistent one was that the 24.3-hour flight had been prematurely called down because Feoktistov had become too ill to continue. There was no evidence, however, then or now, that the flight had originally been scheduled for longer than 24 hours or that Feoktistov had suffered anything more severe than spatial disorientation in which he had the illusion that he was crouching face-down.

Putting a physician in space, was, for the Russians, almost as logical a progression as orbiting a woman. In addition, their flight tested a commodious new spaceship that flouted the previously widely overestimated danger of meteorite hits. "The danger to spaceships from meteorites," says the president of the Soviet Academy of Sciences, Mstislav Keldysh, "is no greater than, say, the probability of a brick falling on your head as you walk along the street." That remote possibility, however, had been taken into account in the Voskhod design, which was specially constructed to confine to a small area any leakage that might occur. The Russians also substantially proved the advantage of a shirtsleeve environment and the considerable advantage of a group flight. Spacemen flying together, for instance, not only showed less emotional strain (one man slept while two worked) but were also able to make joint comparisons of observed phenomena—such as space "fireflies" outside their craft, on which they made separate notes. Feoktistov believed them to be tiny specks of dust emanating from the spaceship itself. Medical tests made on a group also bore more validity and objectivity, and the thoroughness of the medical tests conducted on Voskhod took maximum advantage of the first space-crew subjects. There were also isolated scientific inputs, such as visual and film observations of the phenomena known as air glow and of the aurora polaris. They saw for the first time in the earth's shadow an "upper layer of brightness illuminated by the moon." The feasibility of using a sextant for astronavigation was established as Feoktistov accurately measured the elevation of stars above the physical horizon. "This means," he later reported, "that in future flights of spaceships, particularly in interplanetary flights, it will be possible to determine independently from aboard the spaceship the ship's position, make calculations on the trajectory of the ship's flight, and introduce the necessary corrections."

Of minor significance was the new television system aboard, which permitted ground observation of the view from the spaceship, as well as interior observation of external parts of the spaceship, giving the crew, as Feoktistov called it, "another port." Voskhod also carried scientific experiments involving fruit flies (Drosophila) and living plants, which were monitored for their production of environmental gases. This was yet another evidence of the Soviet's long and intense interest in closed ecological systems.

Despite its solid engineering triumph, the Soviet people's reaction to the flight of *Voskhod 1* was highly enthusiastic but somewhat less than the national outpouring that followed the Tereshkova flight. The Russian man-in-the-street, like his U.S. counterpart, was showing the first signs of a lessening absorption in manned space flight. The cause was a combination of space-age acclimatization and overexposure at the political and press level.

In this atmosphere, and largely oblivious to it, Korolev and company prepared a new Voskhod for flight, at the same time that Red Air Force Lieutenant General Nikolai Kamanin and his cosmonaut instructors put two new space pilots through highly specialized training.

First of all, the Voskhod spaceship had to undergo extensive modification for the inclusion of a bulky and complex air lock that would permit, for the first time, cosmonaut egress into the ocean of space. The Soviet designers knew then that Gemini was being built to allow an astronaut on pure oxygen simply to depressurize his spacecraft, open the hatch, and step outside in a pressure suit linked and sustained by an umbilical cord. But the Soviets did not trust this system, and in addition, they had long ago made the decision for a nitrogen-constituent breathing mixture, which prohibited direct egress for a cosmonaut until he rid his body of nitrogen gas. Otherwise, he would suffer bends as nitrogen bubbles formed in his tissues.

There was still another reason for adoption of the air-lock transfer tunnel. The Soviets were looking beyond the first "space walk" to the orbiting of large crews, some of whom would be required to leave the ship in a vacuum, for either assembly tasks or lunar exploration. If they had adopted the Gemini principle, it would have meant that every crew member—whether or not he went outside—would have to wear a space suit to survive cabin depressurization (or dehermetization, as the Soviets call it). And in Korolev's view, this would create an unnecessary hazard to *all* crew members. Because of the experience of *Voskhod 1*, Soviet designers did not now consider a space suit a very efficient working dress for the cosmonaut. They considered the shirtsleeve environment vastly superior for crew comfort and efficiency.

"If dehermetization and exit by simply opening the hatch was the best variant," said General Kamanin, "we would be the first to adopt it. However, [our] experience showed that this by no means is the best variant. It is not very useful for cosmonautics. The idea was not simply to open the hatch and jump into free space. The exit [air lock] will also be necessary for going over into another ship, or, say, do some work on repairing the ship or coupling one ship with another. . . . Why it will be meaningless to keep all the members of the crew in space suits in a dehermetized vehicle. It is more expedient to have one or two cosmonauts in space suits, put them in the lock chamber, and close the hatch. The other members of the crew may stay in the cabin and work under the unconstrained comfortable conditions. . . . The lock chamber method is much more difficult. But it is more necessary for the development of interplanetary travel."

Selection of the cosmonaut to become the first human satellite of the earth was just as intensive a process as the original selection of Yuri Gagarin. The responsibility and honor fell to a 30-year-old pilot-artist whom the cosmonauts call "the hairy one," because of the abundance of reddish-blond hair on his chest and back. Alexei Arkhipovich Leonov, like Gherman Titov, was born—the eighth in a family of nine—in the Altai region. He spent his childhood on the Baltic and nearly decided to devote his career to the sea, which he loved as a youth. A precocious, talented boy, Leonov early developed his gift of drawing and painting. His mother, now 73, recalls saving his classroom copies of the Russian artists Aivazovsky and Shishkin. Due to the influence of one of his numerous brothers, an aviation technician, Leonov entered the primary Chuguyevskoye Aviation School in the Ukraine when he was 19. Four years later, he had become a skilled fighter pilot. He was selected for the cosmonaut program, like Vladimir Komarov, from the Zhukovsky Air Force Engineering Academy.

During his long training and wait for a spaceflight assignment, Leonov wrote, designed, and illustrated the one-page cosmonaut satiric newspaper, *Neptune*. His other hobbies were the theater and motion-picture photography. He is married to a former medical student, Svetlana; they have one daughter, Vika. One of Leonov's main intellectual absorptions was the moon. He knew it and followed its unfolding secrets with the dedication of a scholar; he made no bones about the fact that he wanted, above all, to become the first man to set foot on the lunar surface. When Leonov was told he was to be the first man to attempt to "walk in space," his first reaction was that it fitted in perfectly with his ambition to be the first to walk on the moon.

Selected as Leonov's flying mate and as the *Voskhod 2* commander was a quiet, reserved, experienced, 39-year-old Air Force colonel, Pavel Ivanovich Belyayev. As a youth of 12 in the thickly forested Vologda region northeast of Moscow, Belyayev was permitted to hunt both alone and with adults. He was a crack rifle shot, and his first ambition was to become a hunter or explorer. He tried to enlist for World War II service at the age of 16, but was rejected because of his age. He later entered flying school from his job as a lathe operator in a factory at Kamensk-Uralsk. His first close call came during a flight with an instructor, when their plane flipped over on landing. He got his wings the same month the European war ended and reported for fighter-pilot duty in the Pacific, where he stayed for eleven years.

Once, on an overwater flight, his gas pump broke and he had to pump gas with one hand throughout the long flight. As a result his hand was numb for several days thereafter.

After he entered cosmonaut training, he, like Vladimir Komarov, was nearly eliminated, and was helped by Yuri

Gagarin to reinstate himself. During a bad parachute landing, Belyayev broke his ankle against a rock. The break was a painful and complex one, and it was widely believed he would not be able to return to the program. Gagarin, however, encouraged him. Belyayev kept lead weights hidden under his bed in order to exercise his pained and weakened ankle. As he rested, he studied his texts on theory. Six months later, he showed up at the gymnasium on crutches. When he felt he was fit enough to jump again, the doctors were against it, but consented when Gagarin agreed to make the critical jump with Belyayev. The jump was a success, and after the lapse of a full year, he was able to re-enter the program.

He impressed General Kamanin one day when Belyayev was inside the chamber of horrors undergoing one of his numerous isolation tests. When an electrical short circuit started a fire inside the chamber, Belyayev, instead of ringing for help or making an emergency exit, promptly put out the fire, patched up the wires, and continued his test. The oxygen-nitrogen atmosphere, of course, prohibited the fire from becoming a disaster.

The Russians, who have a favorite saying, "Speech is silver; silence is gold," seem to have a preference for taciturn men in their space program. Pavel Belyayev, like Valery Bykovsky, Konstantin Feoktistov, and to a certain extent, Yuri Gagarin, was a man of very few words. Partly because of his reserve, the other cosmonauts always addressed him by his full formal name instead of a more familiar nickname. "I find speaking rather difficult," Belyayev admits, "and, in addition, I don't like talking. Perhaps this is because I am not a good talker."

Pavel Belyayev, however, got along fine with the hairy one, Alexei Leonov, and with Komarov and Feoktistov, with whom he spent a great deal of time in specific training for the flight of *Voskhod 2*.

Komarov usually was with the two new crew members as they worked in special simulators, and he often went with them to their actual *Voskhod*, which they also used extensively for training, especially in operating the chamber lock.

A portion of the chamber lock extruded from the cabin, and its other end used up part of the third-seat space inside the cabin. The separate instrument panel for the chamber was located within easy reach of both seats. Inside the chamber were frosted bulbs for illumination and a separate system of controls so it could be operated from inside. In addition to the television cameras overhead in *Voskhod*'s cabin, a camera was located inside the chamber and a second one was installed outside the chamber exit into space.

Leonov got the most intensive and specialized training of any cosmonaut. Soviet psychiatrists were seriously concerned about what they called "the psychological barrier" they felt would confront a man swimming alone in space for the first time. They did not, however, reveal to Leonov the full depth of their concern, but he must have suspected it from the numerous psychological tests they subjected him to. Color movies of him in extended isolation training show him, completely unselfconscious, sitting in his undershirt calmly sketching boats, trees, and houses. Leonov has a chest like a gorilla, and from time to time he flexed his powerful and hairy arms as if mentally rehearsing his swim in space.

Toward the end of their training, Boris Yegorov and other doctors rehearsed both crew members on the vigorous schedule of medical tests designed to extend and enlarge the significant data obtained on the preceding troika flight. Like *Voskhod 1*, *Voskhod 2* was programmed for 24 hours, and the work time of both men was fully scheduled. The vestibular, optical, and other medical apparatus that had flown on *Voskhod 1* was also included in *Voskhod 2*. In addition, there were new psychological tests that were designed to study the effect of weightlessness on memory. Feoktistov, who had done the filming on the previous flight, gave them both lessons in the specialized camera techniques required beyond the atmosphere.

A specific part of Belyayev's training was in direct preparation for a manned landing on the moon. His experiments in astronavigation, continuing those of Feoktistov, were designed to make position, trajectory, and speed calculations independent of the existing ground computing centers. This development of an autonomous navigation system for Voskhod was designed to permit the spaceship commander eventually to make orbit and lunar trajectory corrections and to determine accurately the instant of switching on the new retrorocket in order to land at a predetermined area on the moon.

To prepare Leonov for egress through the air lock, a Voskhod replica, complete with air lock, was installed inside a transport aircraft. As the plane flew its weightless parabolic arc, Leonov repeatedly practiced entering and leaving and working with the slightly stiffened umbilical, or tether. For these exercises, he wore the new self-contained space suit designed to afford protection against vacuum, infrared, temperature extremes, and visible and ultraviolet solar radiation. On his back he wore a boxlike, rather cumbersome but effective life-support unit. His white helmet had a dark visor filter operated by a simple mechanical lever.

In a noteworthy and sensible modification to previous cosmonaut-designer relationship, Sergei Korolev now insisted that cosmonauts participate directly in all design and modification decisions. Doubtless his decision grew out of the practical Vostok modifications suggested by Gherman Titov and other cosmonauts after they examined the virtually completed spaceship.

Both Belyayev and Leonov and others, especially Feoktistov, participated in the development and testing of all new systems and equipment. Says Leonov: "We were present at all tests and introduced the changes that we thought necessary. We were happy to see that the designers did not leave a single suggestion of ours without notice. . . . The tester is an important figure, of course. Nevertheless, we tested all the new units ourselves."

This new practice undoubtedly contributed enormously to the confidence with which Leonov and Belyayev ascended to the top of their rocket on the morning of March 18, 1965. The same procedures and launch-pad traditions of farewell and formal speeches of previous flights were followed. The most noticeable difference at the pad was the presence of late winter snow, some of which was banked up around the huge flame deflector at the rocket's base.

At precisely 10 A.M., the nozzles discharged their powerful jets of fire, which turned the snow into instant steam. What follows is a report of a portion of the historic flight events in the words of its formerly reticent commander and his space-walking co-pilot:

BELYAYEV: The spaceship was put into an orbit with the following parameters: the minimum distance from the earth's surface [perigee] was 173 kilometers [108 statute miles] and the maximum distance [apogee] 495 kilometers [308 statute miles]. The orbit's angle of inclination was approximately 65 degrees and the period of revolution of the spaceship around the earth was 90.9 minutes. All the parameters of the orbit were close to those calculated.

Our assignment envisaged a flight along the orbit of an artificial earth satellite with the experiment of emerging from the spaceship into outer space in the process of the orbital flight. If I were to be asked whether all this was easy or difficult, my reply would be that this was no easy job. I think that Alexei Leonov will confirm this.

The experiment involving man's emergence into outer space was started actually as soon as we had gone into orbit. We felt very fit after the boost stage. The entire cycle of operations fully coincided with those mastered on earth, the only additional thing being weightlessness, which in the given case even helped to carry out the assignment.

My friend Alexei Leonov conducted all the preparatory operations for emerging into outer space with high precision and eagerly awaited my order. He wanted to slip out into outer space even earlier than the time envisaged by the program, but I restrained him. A program is a program, and as the commander of the crew I was responsible for its fulfillment.

LEONOV: While in the spaceship's cabin I put on, with the skipper's help, a haversack-type unit with the autonomous life-sustaining system and began using it. I did this before going out into the lock chamber.

We leveled out the pressure in the cabin and in the lock chamber. Then we opened the hatch from the cabin into the lock chamber, and it was through that hatch that I swam into the chamber.

I brought the pressure in the space suit to a set level, checked whether it was airtight, saw to it that the helmet was closed properly and that the light filter on it was in the right position.

Having checked the supply of oxygen in the space suit and gone over all the operations of stepping out of the spaceship in my mind's eye, I prepared to sally forth into outer space.

Belyayev closed the hatch of the cabin and drained the pressure from the lock chamber. The skipper opened the exit hatch, and a blinding ray of sunlight filled the chamber—it was as though someone was doing some electric arc-welding.

I leaned out of the hatch and almost went right out. "I'm pushing off," I said to Pavel. "Not so fast," he said. "Wait a tick." I stayed there half out of the hatch and waited. The way into the void of space was open. I was impatient.

BELYAYEV: Having satisfied myself that all of Alexei Leonov's life-sustaining systems were functioning normally, that his pulse and respiration were normal, at the prescribed time I gave him the order to emerge into outer space.

LEONOV: At last, everything was ready and I could emerge. I stuck my head out of the exit hatch.

The boundless expanses of outer space unfolded before me in their indescribable beauty. I took my first look at the earth. It sailed majestically before my eyes and seemed flat. It was only the curvilinear shape of the edges that reminded me that it was a globe.

In spite of a pretty dense light filter, I saw bright clouds, the azure of the Black Sea, the fringe of the coastline, the Caucasian Range and the Novorossiisk harbor.

The moment came to leave the ship and step out into space, the moment for which we had prepared for so long, of which we had thought so much. Unhurriedly, I climbed out of the hatch, pushed myself away from it gently, moving further and further away from the ship.

BELYAYEV: I clearly felt everything that Alexei was doing to the ship. I felt it when he pushed, and heard his boot ring as it scraped against the side of the spaceship. And even when he ran his hand over the side of the craft, I heard a kind of rustle.

LEONOV: As I pushed off, I felt as though the ship bounced away in the opposite direction. According to the laws of mechanics, that was how it should have been. But the sensation was unfamiliar. Probably it was more like being on a swing than anything.

The lifeline that connected me with the ship stretched to its full length and my movement away from the ship ceased. The slight effort I had exerted in detaching myself from the ship had imparted a slight angular movement to it, and a full view of our wonderful spacecraft slowly unfolded before my eyes.

I expected to see sharp contrasts of light and shadow, but there was nothing of this kind. The parts of the ship in the shadow were illuminated well enough by the sun's rays reflected from the earth.

Swimming in space is nothing like swimming in wa-

A demonstration in Red Square on June 22, 1963, welcoming Valery Bykovsky and Valentina Tereshkova after their dual flight in *Vostok 5* and *Vostok 6* during which they came within three miles of each other. This enthusiastic demonstration marked what many observers believe to be the apex of Soviet "rocket fever".

Yuri Gagarin

Valentina Tereshkova

Alexei Leonov

Vladimir Komarov

Sergei Pavlovich Korolev (1908-1966), brilliant chief spacecraft designer, winner of the Lenin Prize, and spiritual head of the early Soviet space programme. The first manned flight undertaken without his direction resulted in the tragedy in which Cosmonaut Colonel Vladimir Komarov, who had earlier been interviewed by the author, was killed.

The funeral ceremonies for Korolev at the Kremlin Wall, January, 1966.

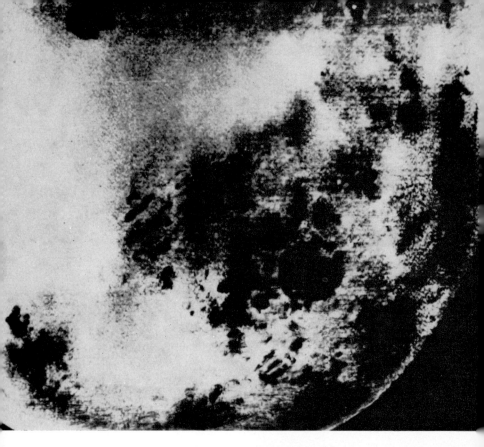

Luna 3 and one of the rudimentary photographs radioed to earth providing man's first view of the dark side of the moon in October, 1959, the second anniversary of *Sputnik 1*.

ter. In the latter, you feel the water streaming around your body, feel the resistance of the medium. And you have to keep your body in a more or less definite position.

In space, you can float about as you like. I, for example, stretched out my arms and legs and soared. And it's more convenient. There's plenty of room to move in. You breathe easily, even better than on earth. It is true that my pressure suit resisted changes in the form of my body, flexing of arms and legs, and so it required an effort to work.

I pulled the lifeline slightly and started slowly moving toward the spaceship. When I reached it, I pushed myself away again and began to move gradually away from the spaceship, turning about my transversal axis.

I saw the universe in all its grandeur. The view of untwinkling stars on a dark violet background changing to the velvet black of the abysmal sky was followed by views of the earth. I saw floating before me great, green tracts of land. I recognized the Volga, the mountain range of the hoary Urals. Then I saw the Ob and the Yenisei rivers, as though I were swimming over a vast, colorful map.

As usual, there were clouds over the mountains. But over the coast itself, there was wonderful, sunny weather. Everything was very clear. No mist. I could even see streams and ravines.

The great distance made it impossible for me to identify towns and the details of the relief. However, for those who are familiar with the brush and easel, it would be difficult to find a more majestic panorama than the one I beheld.

BELYAYEV: During the period when Leonov was in outer space, I followed all his actions, using a television installation for the purpose. The TV system operated very well and clearly. I saw in detail all the actions conducted in outer space, and even distinguished details of his space suit. My phone and the instruments in the cabin made it possible for me to check the operation of his individual life-sustaining system, as well as his pulse and respiration.

I thus followed Leonov's actions and his condition continuously and in all respects while he was in outer space. I must say that he had been well-trained physically on earth, and this insured the successful fulfillment of the task of emerging into outer space. During the most strenuous operations, requiring physical effort, his maximum pulse was 135 beats per minute.

The spaceship's system made it possible for me, as the ship's commander, in case of necessity, to come to the assistance of the cosmonaut in outer space. However, no such necessity arose.

LEONOV: I acted as though I was in familiar circumstances. One of my first jobs was to take the cover off the camera that had to photograph me. I held the cover in my hand and thought to myself: Shall I launch it into another orbit? But I decided not to litter up the cosmos. I heaved it earthward with all my might, and watched it until it disappeared from view.

Some time later, I made a pretty strong pull at the lifeline and was forced to use my hands to keep clear of the spaceship which started moving swiftly toward me.

My first thought was not to strike the spaceship with my helmet. So, as I flew up to the hatch, I softened the blow with my hands. This proved very easy to do, and I saw that with enough training one could move fairly efficiently and with good coordination in those unusual conditions.

I felt fine, was in excellent spirits and did not want

to part with free space. And even after I had received the order to return to the spaceship, I pushed myself away from the hatch once more to check the origin of the angular velocity in the first moment after the push. As to the so-called psychological barrier, which was supposed to be an insurmountable barrier to a man about to meet the voids of space face to face, I must say that I did not feel any barrier at all and even forgot that such a barrier might exist. There was no time to think about it.

After all, the 20 minutes that I spent directly in outer space, including the 10 minutes outside of the spaceship, were the "highlight" of the *Voskhod 2* flight. I understood this very well and therefore did everything I could so as not to waste a second. Unfortunately, time flew by very quickly, and the last moments of my stay outside the ship arrived. I removed the cinecamera that had recorded on film my sortie into space, and tried to get into the hatch immediately.

But this proved not so easy to do. Firstly, I had to exert more effort, and I was a little tired. Despite everything, movements in an inflated space suit are somewhat restricted. I had to make quite considerable physical efforts, and my farewell with space was a bit drawn-out. I was supposed to wind the cable around my hand and haul myself in. I saw that it was going to be a long, drawn-out affair. I thought a bit, mentally pulled myself together and hit on a way of doing it quicker.

The Volga was very clear—probably I was turning a somersault at that moment. Then I saw the Irtysh and the Yenisei Rivers. That was the last I saw. After that, I went back into the cabin. The C.O. quickly shut the hatch behind me and turned on the pressure.

BELYAYEV: I greeted my colleague with "Good for you!" Then Alexei had a little rest and got down to work.

He had to write all his impressions down in the log-book immediately. He wrote for about an hour and a half.

LEONOV: When I was already sitting in my seat, I felt streams of perspiration running down my forehead and cheeks. I think that it is a bit too early to compare outer space with a place for an entertaining stroll, as some journalists do. I am sure that, without the many months of all-around training, of which much has been said, I would never have managed to cope so easily with the task set me.

BELYAYEV: After carrying out the experiment of the exit into outer space, we conducted a series of scientific observations and investigations, and took many films and photographs while continuing on orbital flight. During the entire flight, all the systems and equipment of the *Voskhod 2* functioned normally. The temperature within the cabin was approximately 18° C; the humidity, 35–40 percent; and the pressure, 1 atmosphere. For greater convenience in conducting certain scientific experiments and observations, we removed parts of the space suit, and in particular the helmet, boots, and gloves and, it should be said, we felt fine.

We violated the program as far as sleeping was concerned. The thing is that we had a lot of interesting work and our train of thought was as follows: sleep is a good thing when you want to rest, but we had not taken off for outer space to have a rest. We had done this to work, and so we should work. On the other hand, the doctors might be upset. After all, it is important for them to know how a person sleeps in outer space. I hope they will forgive us this time.

According to the flight program, we were to land during the 17th circuit employing the automatic cycle of descent and using the system of solar orientation.

In the process of preparing for a landing according to the automatic cycle of descent, we noted certain abnormalities in the functioning of the solar orientation system. This actually delighted us. We had the opportunity of landing under manual control and thus demonstrate another remarkable feature of Soviet piloted spaceships—piloted, it could now be said, in the full sense of the word. Frankly speaking, we were only afraid that we would not be allowed to do this. After all, the automatic descent system could be used on the next circuit. A reply to our request for a manually controlled landing seemed to drag on and on. Finally, we received permission for a manually controlled landing on the 18th circuit. Ground control was confident of our ability and was certain that we would cope with the task. I had all the necessary calculations, which had been worked out on the earth. In accordance with those calculations, I switched on the retro-engines.

As a fighter pilot, I had made no small number of landings in modern, high-speed planes. However, the speed of our spaceship *Voskhod 2* could not be compared with the speed of the most modern plane. And it was necessary to land such a super-speed vehicle by manual control—moreover, in the required area.

The manual control system operated faultlessly and we landed approximately where we expected to land—although we somewhat overflew our mark due to the novelty of such a landing. The spaceship came down between two big fir trees. It landed in deep snow and began to settle under the burden of its own weight. We opened the hatch and looked around. Everywhere there were huge trees. The snow was thick—from one and a half to three meters, and on its surface were the tracks of hare and fox. Within a few minutes, we reported by radio: "We have made a good landing."

Any landing of a spaceship, as of an airplane, is a good one if you can walk away from it. In manually landing *Voskhod* for the first time, however, Belyayev missed the designated landing area, like Scott Carpenter, by a considerable margin. The overshoot caused them to come down in an area that previous spaceships had been programmed to avoid—the towering Ural Mountains of west central Russia. It was the first truly emergency landing of any spacecraft that called upon extensive cosmonaut training in all-climate survival. As Belyayev indicated, *Voskhod* plunked down in deep snow in a remote section of a dense forest the Russians call the taiga—birch and fir mixed. The location was in the general vicinity of the town of Perm on the eastern slopes of the Urals.

Not mentioned in the above report by the space pilots was the fact that it was nearly two days before a rescue party could reach them and nearly four days before they were able to return to Tyuratam. The first night spent in the forest was one of bitter cold accentuated by the acute discomfort of medical sensors still attached to their bodies. The next day, in response to their radio pleas, search planes dropped warmer clothing and extra survival gear. Fortunately, the weather was clear and a ground rescue party, including a doctor and a movie cameraman, reached them before they had suffered anything worse than acute discomfort. According to the film documentary, the first thing the doctor insisted they do was to strip and take a snow bath, which they did in gleeful spirits. After donning fur boots, warm clothes, and heavy green jackets with fur collars, the cosmonauts ate a hot campfire meal and responded to the first request for their autographs. Then the party set off down the steep, forested slope on skis. After an exhausting journey of nearly twelve miles, they arrived at a comparatively level area, where lumberjacks toppled enough fir trees to create a small, square helicopter landing pad. The chopper hoisted them up out of the taiga and took them to the nearest airport, where jubilant Ural mountaineers welcomed them. A plane then flew them to Tyuratam, then an Ilyushin-18 airliner flew them to Moscow with the traditional escort of seven MIG jets in tight formation.

By the time the two men walked the long red airport carpet to receive the traditional adulation, speeches, and flowers, they were fully recovered. In the Moscow parade, they waved from an open-topped limousine completely bedecked in flowers designed in the shape of a diamond, which was their *Voskhod 2* call sign. Russia once again went wild over a successful manned space flight.

Also not mentioned in the above initial report was the fact they they both unexpectedly saw a U.S. satellite in space. "The satellite was in an orbit which was somewhat higher than the orbit of our ship," Belyayev explained later. "It lagged behind and therefore there could be no collision. Nevertheless, the possibility of collisions in space when there are several ships there is not ruled out, though the practical probability of this is very small."

Leonov, like the other cosmonauts, showed some concern over the growing amount of "space garbage," which he feels will ultimately make orbital space dangerous for man. Elaborating on his throwing away his camera, he explained: "I thought about the possibility of littering up space when I held it in my hand. Of course, this thing would not have done much harm because it is small and has no sharp ends. But still it would have been hanging about in space for a very long time. Therefore, when I threw it from me, I directed it toward the dense layers of the atmosphere so that it might have burned there. I think that international organizations and the countries launching satellites must work out a kind of space dispatching service to prescribe the orbits of satellite ships or even manned spaceships. This will be quite important."

The key medical investigations of *Voskhod 2*, of course, were those pertaining to Leonov during his individual excursion in space. The EKG sensors and the ohmic transmitter for measuring his chest perimeter that were taped to his body led, first of all, through waterproof circuitry to a special optic indicator that Belyayev could read inside the cabin. Then, via telemetry, they fed to the medical consoles in the underground control center. As Leonov prepared for exit, his neuropsychic tension gave him a pulse rate of 90–100 beats per minute. During his ten-minute period outside, during which he was occupied with a number of working operations, his pulse varied between 147 and 162 beats per minute. During this time, of course, he was weightless, but he was having to work against the considerable obstacle of his pressure suit, a demand U.S. astronauts later found surprisingly and exceedingly fatiguing. (Astronaut Ed White's pulse rate, however, never rose higher than 130 beats per minute.) By the third revolution, the pulse rate of both cosmonauts had settled down. Belyayev's, which had risen to a high of 130 just before he opened the lock, returned to exactly normal; Leonov's actually settled below normal. Belyayev's second high point was reached during his 17th-orbit stress of manually positioning *Voskhod 2* for re-entry. There was no spatial disorientation of either cosmonaut. Each lost the same amount of weight, slightly over one pound. According to a detailed study of a group of seven Soviet doctors, Leonov's ten-minute "walk" resulted in "no severe changes in the state of the basic functional system of the organism and no noticeable reduction of the level of work performance."

There were also other medical experiments inside *Voskhod 2*. In the first multiple-crewed flight without a doctor aboard, Belyayev and Leonov performed experiments on each other much as Yegorov had done with his two fellow crew members. These were chiefly vestibular experiments and blood tests, both made while the spaceship was stabilized and non-orientated. They also used tables investigating the light sensitivity and resolution capacity of the eyes as a special tracking device followed all movements. To investigate the gaseous exchange in weightless organisms, they took samples of exhaled air, which they brought back to earth.

In the absence of any negative physiological or psychological factors affecting a human satellite of the earth, the Soviet medical investigative team cautiously concluded that the flight of *Voskhod 2* "opens new prospects for man's mastering of outer space."

The Russians had thus far orbited, in addition to jet pilots, a physician, a scientist, and a woman textile worker. In

Alexei Leonov, they had also orbited an artist, and the talents of a painter were not as extraneous in space as they might first appear. I studied Leonov's space paintings in Moscow, and they illuminate certain areas that had so far not responded well to cosmonaut photography—particularly the bright color bands of earth. As Leonov explains it, "previous cosmonauts tried to photograph three varied belts of light—red, orange and blue—that ring the earth, but they never showed up in photographs." Leonov's drawings and paintings, some of which were sketched in space, were representational and as literal as he could make them. One painting shows the three bright light bands of earth against the inky blue-blackness of space as a winged and dazzling sun peeks out between the hued strata of light—like a flashlight bulb cupped in the blood-pulsing fingers. Other paintings are more fanciful, such as one of two earthmen—one undoubtedly the painter himself—standing on the surface of the moon. Another shows the fiery and soundless blast of the *Luna 9* retro-rocket as it brakes its hurtling descent to the moon. Like Gherman Titov, Alexei Leonov is a man of acute perceptions and keen sensibilities. It is a credit to the discipline of both their mentalities that they survived the regimen and intensity of their training to use their individual talents to illuminate the new and very special world of the cosmos.

13
Automatic Earth Satellites

IN THE decade following the first successful Sputniks, Russia, like the U.S., populated near-earth space with a great variety of large and small unmanned satellites. In addition to manned spacecraft and lunar and solar-system probes, by 1967 they had put into orbit approximately 150 satellites, including 46 that were still in flight. These included scientific satellites for continuous atmospheric studies, communications satellites, military photographic and reconnaissance satellites, biological satellites, maneuverable satellites, navigational satellites, geodetic satellites, and nuclear-blast-policing satellites.

One of the most interesting aspects of the unmanned space-flight program of the Soviet Union is the virtual absence in its scientific literature of the running debate that has taken place in the U.S. on the relative merits of manned and unmanned flights. The U.S. dichotomy between the "black box" advocates of California's Jet Propulsion Laboratory and other groups at odds with the "warm body" advocates of NASA and of the Manned Spacecraft Center in Houston seems to be an indigenous phenomenon, probably based on our particular method of chopping up the Federal budget. The paradox is interesting because the Soviets, from the beginning, have had to make innumerable decisions on when to send manned and when to send instrumented satellites and probes into space. Vladimir Pelipeyko, writing in the excellent Soviet publication *Nauka i Tekhnika* (*Science and Technology*) is one of the few who have discussed this decision-making process: "... if the control of a spaceship is entrusted completely to automatic apparatus it will have one substantial shortcoming—a low reliability. As shown by computations, the reliability of an automatic control system of a ship designed for flight around the moon and return to earth is 22%. However, with the participation of man it increases to 70%."

Pelipeyko admits that, "The principal role in the Soviet study of space is being played by different kinds of cybernetic equipment," but at the same time he affirms that the "creative activity of man in space . . . makes it possible to formulate more perfect programs." Unlike a number of U.S. spokesmen, however, Pelipeyko seems to favor what he calls "an integrated man-machine system . . . there can be no talk of the complete replacement of man by cybernetic apparatus."

However the Soviet inventors and designers regard their cybernetic and automated data gatherers, they have made widespread and ingenious use of them across virtually the entire spectrum of astronautics. We shall consider here only those whose application is not primarily military; a later chapter will deal with all military applications.

On March 16, 1962, Russia launched *Comos 1*, the first of a new and vaguely defined series of satellites whose purpose, the Soviets said, was scientific investigation. *Cosmos 1* was to study the earth's radiation belts and to check possibilities for long-range radio communications. By July 28, 1962, the Russians had launched seven of these new "scientific" satellites. We were to discover later, however, that the Cosmos group of satellites had many functions, many of them military or quasi-military. Most U.S. observers estimate that approximately 50 per cent of the nearly 150 Cosmos satellites launched in the first decade were military in nature. Those that were not primarily military had more or less the same configuration: a short cylindrical section capped by hemispheres spiked with antennas at each end.

For a number of months, Moscow correspondents dutifully reported the regular, unelaborated Soviet announcements of Cosmos satellites, and U.S. newspapers dutifully repeated them, sometimes on the front page. But it soon became apparent that the Soviets were simply not elaborating on this series and, perhaps as the Soviets intended, the Western press finally relegated the brief Cosmos announcements to the status of English football scores. Cosmos announcements finally became so routine that not even *The New York Times*, "the newspaper of record," always reported them. The real news that the scientific Cosmos satellites were making was in the esoteric reports of Soviet scientific journals, but these went largely unread and uninterpreted.

In the same year as the second Soviet near-rendezvous manned flight (1963), the Russians also launched a new satellite (of about the same weight as Vostok) having powers of maneuverability.

Polyet 1 (pronounced *polyot*), the world's first truly maneuverable satellite, was launched from Tyuratam on November 1, 1963, amid a surprising amount of high-level cognition, considering that it was an unmanned shot. In Moscow at the reception for Laotian Prime Minister, Souvanna Phouma, Khrushchev said:

> I will not attempt to hide my happiness in connection with one very pleasing piece of information which I would like to tell you: Today the scientists of the Soviet Union have launched a new spaceship "Polyet 1." The fact that we are launching spaceships is not new. For several years now we have been engaged in launching spaceships and these are safely traveling through space today. However, the spaceship that we have launched today is entirely new. If, in the past, spacecraft which were injected into orbit have primarily carried out their flights in the direction into which they have been launched, the new spaceship which we have launched today is carrying out broad maneuvers in space changing its orbital plane and altitude.
>
> The fact that we have launched such a spacecraft is proof that the human brain has risen to a higher level. Now a man in space is not any more chained down to

his spacecraft. The man controls the spacecraft and guides its flight. The spacecraft becomes more and more obedient to the will of man.

Here is another area which opens unlimited possibilities for peaceful competition between nations. The purpose of such a competition is to launch spacecraft for peaceful purposes and to utilize them for the conquest of the universe which was not studied by mankind yet....

I propose a toast for the triumph of the human mind, for science and for the scientists who have created such a spacecraft as "Polyet 1" and have launched it successfully. This spacecraft is now in flight.

I propose a toast to the President of the Soviet Academy of Sciences Keldysh, to all our remarkable scientists, engineers, technicians, and workers, who have through their efforts created our spaceships.

U.S. tracking facilities showed that *Polyet 1* could indeed change altitude. Its initial orbit had an apogee of 329 miles and a perigee of 211 miles; after rocket motors were started in space, the high point of orbit became 892 miles and the low point remained about the same, 210 miles. *Polyet 1* also made minor changes in its orbital plane from its original launch inclination of 58° 15'.

One of the intriguing questions of the Soviet space program is why the maneuver capability of *Polyet 1* was not then incorporated in the flights of *Vostok 5* and *6* in order to achieve full rendezvous and docking. The enigma arises from the denial of heretofore invulnerable Soviet logic in space scheduling.

After the initial fact of the launch, however, Soviet spokesmen did begin to construct a new kind of space future based upon the demonstrated maneuverability in space. The official news agency, Tass, said Polyet could be used "for taking pictures of the earth and its cloud cover and as a relay station for transmitting radio and television broadcasts."

Izvestia said the new satellite would lead to the assembly of a platform in space. Soviet scientist Professor Georgi Pokrovskyis commented accurately and prophetically: "It will be possible to fly in such a way that the ship is suspended without motion in the zenith over any points of earth, to fly along a circular orbit whose center can be switched from the center of earth to any dimension." Pokrovskyis, a physicist at the Zhukovsky Air Force Engineering Academy, suggested a heavy fuel supply aboard future Polyets. He spoke of a $60°$ turn that would require the same amount of fuel as that required for inserting Polyet into orbit. He also suggested future reversal of orbital direction, requiring double the launch fuel. Other spokesmen commented on Polyet's implied ability to steer itself around dangerous radiation zones somewhat as a jetliner avoids a cumulonimbus cloud.

In view of the wealth of capabilities and virtues attributed to *Polyet 1*, the launch of *Polyet 2* on Cosmonaut Day, April 12, 1964, was greeted with considerable expectations. But the world was to experience a long delay before the Soviets made substantial *use* of the maneuver capability demonstrated by the Polyets. *Polyet 2* did carry out repeated maneuvers during its initial orbits, but these were only environmental engineering tests. It soon became apparent that *Polyet 2* had no *functional* task in space involving its ability to maneuver.

Again Soviet spokesmen commented on the use of Polyets in assembling space stations and ferrying up and releasing orbiting crews. But a long and inexplicable gap was to follow these first two maneuverable satellites. The delay stretched over three years, and as it increased some U.S. observers took a second look at the basic purpose of the first two Polyet flights. To the Director of Foreign Technology Studies of California's Data Dynamics, Inc., Joseph L. Zygielbaum, the maneuverable satellites may have been the direct forerunner of the payload of what the Soviets later called their global rockets. Khrushchev had described a Russian orbital bomb as capable of orbiting the earth and then "striking by command any given target on the earth's surface. If this is true," suggested Zygielbaum, "such a vehicle or weapon should have a capability to maneuver freely in space. . . . It is quite possible that this weapon is a result of the experiments with the Polyet satellites." This possibility receives some support from the fact that certain Cosmos satellites, especially *Cosmos 22*, have also been observed to maneuver in space—a function about which the Soviets have maintained complete silence.

A second automated-satellite program that Russia inaugurated in 1963 was the highly imaginative Electron program. The initial launch of January 30 was the first time the Soviets launched two orbital payloads from a single booster. Their announced purpose was to study radiation and the magnetic field in the upper atmosphere, but their method of doing this was unique. *Electron 1* swung into a relatively tight orbit close to earth and entirely within the bounds of the inner radiation belt, while its companion, *Electron 2*, swept through a great elongated orbit that carried it at apogee more than 36,000 miles from earth, passing through both the inner and the outer radiation belts of earth. This permitted simultaneous readings on the effect of the sun within various strata and segments of the earth's electromagnetic sheath.

The technique of creating two simultaneous but highly varying orbits was exceedingly complex. "In practice," explained physicist Yu. I. Logachev, "the problem of injecting two satellites into entirely differing orbits by means of one carrier rocket represents considerable technological difficulties. For the purpose of injection of the spacecraft *Electron 1* and *Electron 2* into the desired orbits, it was necessary to execute the separation of the first of these satellites along the active flight sector of the rocket's last stage at a moment when its motor was still in operation. The separation of *Electron 1* must have been accomplished in such a manner so as not to create disturbing moments that would affect the operation of the guidance system of the last rocket stage, as well as the accuracy for injection of *Electron 2*. Precautions had to be taken to prevent *Electron 1* from becoming entrapped in the zone of the jet exhaust of the last rocket stage.

"Both these difficulties were overcome by the application of a special relative system which secured the separation of

Electron 1 from the carrier rocket with a strictly defined velocity. The separation took place without any disturbing effects on the further flight of the last rocket stage. In addition, the structure of *Electron 1* was designed in such a manner that at the moment of separation its dimensions were very compact without any protruding parts on its surface."

The two spacecraft in orbit were quite different. *Electron 1* looked like an ashcan with six solar panels affixed to it like antic fly swatters. *Electron 2* resembled the tapering top of a skyscraper. When Soviet Academician Anatoli Blagonravov told world scientists assembled in Florence, Italy, of the type of instruments contained in the Electrons, there was considerable surprise among those scientists who had previously believed erroneous reports that Soviet astronautical equipment was crude and rudimentary.

Instrumentation on *Electron 1*, as listed by Blagonravov, included:

1. Five scintillation, gas discharge, and semiconductor radiation detectors for studying Van Allen belts. These devices record electrons with energies ranging between 40,000 electron volts and 10,000 electron volts and protons in the energy spectrum ranging from 2 million electron volts to 200 million electron volts.

2. Soft corpuscular radiation detectors consisting of two fluorescent indicators with an electron accelerator on one and a magnet for shielding against low-energy electrons on the other. Energy spectrums investigated are 5,000 electron volts and above for electrons and 150,000 electron volts and higher for protons.

3. Mayak (beacon) system for studying the properties of the ionosphere and interplanetary space by means of coherent radio waves with frequencies of 20.005, 30.007, and 90.0225 mc. This apparatus, also in several Cosmos craft, probably is similar to that used on the Canadian Alouette satellites.

4. Ballistic piezoelectric micrometeorite detectors

AUTOMATIC EARTH SATELLITES 197

sensitive to masses about 10^{-8} grams and having a sensitive surface area of 0.325 sq. ft.

5. Mass spectrometer for studying upper-atmosphere ion composition. Sensitivity span includes ions ranging in weight from 1 to 34 atomic mass units.

6. Solar cell samples for testing under space conditions.

For *Electron 2*, Blagonravov listed the following instrumentation:

1. Radiation belt instruments identical to those on *Electron 1*.

2. Solar cell elements and a mass spectrometer for ion studies, as in *Electron 1*.

3. Electrostatic spherical analyzer registering electrons and protons with energies in bands spanning 30 per cent above and below 200, 400, 1,000, 2,000, 4,000, and 10,000 electron volts.

4. Two ferrosonde magnetometers for measuring all three components of the geomagnetic field in ranges from 2 to 3 gammas to 1,200 gammas for each component.

5. Solar X-ray counters mounted on a sun-oriented platform for measuring X rays with wavelength intervals from 2 to 8 Angstroms (A) and 8 to 18 A.

6. Cerenkov and scintillation counters for studying cosmic-ray composition and time rate of change of the fluxes of different groups of nuclei with energies above 600 million electron volts.

7. Cosmic radio noise recorders operating on frequencies of 725 and 1,525 kc.

8. Charged-particle trap registering positive ions in the earth's plasma envelope and protons of solar corpuscular streams. Trap also detects electrons with ener-

gies at levels greater than 100 electron volts, which cause collector currents of opposite polarity.

Later, in a *Pravda* article, Yu. I. Logachev and two colleagues elaborated on some little-known effects of radiation and the highly sophisticated Electron investigations. Portions of their explanation are quoted here because of the insights provided into the scientific motivation behind the amazing Electrons. This excerpt also clearly summarizes much of what Soviet scientists had learned about satellite operations and characteristics:

> One of the primary missions of the satellites *Electron 1* and *Electron 2* is the study of the internal and external radiation belts of the Earth. The fluxes of charged particles within the radiation belts are extremely large. These particles bombard each object which penetrates the radiation belts. The energy of many particles in the radiation belts is so great that these particles are capable of penetrating a spaceship.
>
> Exposure to this type of radiation is dangerous not only for the well-being of cosmonauts engaged in lengthy travels through the radiation belts, but also due to the fact that radiation is capable of changing the properties of various materials used in the construction of spaceships. It has been established that the quantity of electrical energy which is produced by cesium solar batteries aboard spaceships and cosmic rockets drops when such vehicles are exposed to this type of radiation. In the case of extremely intensive radiation, solar batteries might be destroyed completely. Such was the case with certain American satellites after a sharp increase in radiation intensity within the radiation belts as a result of a high altitude nuclear explosion which the United States set off on July 9th, 1962.
>
> It is also a known fact that certain transparent mate-

rials lose their transparency under the influence of radiation, which can be particularly unpleasant in the case of optical systems. Radiation effect destroys many organic materials which are used in the form of thin films for the purpose of adding various properties to surfaces, as for instance for the translucence of optics.

The study of the behavior of various materials in cosmic space became the main subject of a recently-developed scientific branch which is known as "Space Materialography."

In order to determine the reliability of a given material during a space flight, it is necessary to know the dose of radiation to which such a material will be exposed. For the purpose of prognosis of the radiation dose, scientists must know not only the state of the radiation belts at the given moment, but also they must be able to predict the immediate future state of the radiation belts. It is therefore necessary to know the laws which regulate the radiation belts, as well as to understand the nature of their origin and existence.

A satisfactory explanation on the nature of the Earth's internal radiation belt, which was discovered by American scientists with the aid of the satellite *Explorer 1* is presently available. Nuclei of atoms, which are component parts of the Earth's atmosphere, disintegrate under the effect of cosmic rays. In the process of disintegration of the atomic nuclei, the component particles [neutrons] become dispersed in all directions, some of them leaving the atmosphere. The life span of a neutron is only twelve minutes. New charged particles—protons and electrons—emerge during the decay of neutrons. When a neutron becomes decayed near the Earth, the proton and electron become entrapped by the Earth's magnetic field and begin to travel along spiral trajectories which take them from the northern hemisphere to the southern hemisphere and back again along magnetic force lines. Such a particle completes hundreds of millions of "trips" from one hemisphere to the other before perishing. Each such trip takes less than one second. This circumstance testifies to the fact that the Earth's magnetic field creates a "trap" for charged particles. This trap might accumulate many particles since at high altitudes above the Earth, the density of matter is very low and particles which travel through that region lose their energy in a very slow manner. This hypothesis explains very well the experimental data on the content and energy spectrum of particles within the internal radiation belt. Furthermore, by comparing theory with experience, it is possible to obtain information on the density of the atmosphere at altitudes higher than 1000 kilometers.

An entirely different situation is observed in regard to the external radiation belt which was discovered by Soviet scientists during the flight of the third artificial Earth satellite (*Sputnik 3*). We might consider as an established fact that the above-discussed mechanism of creation of the internal radiation belt cannot be used as an explanation for the existence of the external belt. Therefore, the external radiation belt represents a mystery at the present time. Apparently a unique "cosmic accelerator" of particles is active near the Earth at distances of thousands and tens of thousands of kilometers. On the basis of data obtained during the satellite flights, we know the type of particles which are primarily accelerated by this "acclerator." However, we still do not know the structure of this "accelerator."

When the Earth passes through corpuscular fluxes emitted by the Sun, magnetic storms and polar glows (Aurora Borealis) can be observed. During this period of time, most extreme changes take place within the external radiation belt. This indicates that during this period of time, the "near-Earth cosmic accelerator" is active.

This is why it is necessary to conduct simultaneous investigations of various physical phenomena in order to solve the mysteries of near-Earth cosmic space. This requires the creation of a space system consisting of a number of satellites which would conduct simultaneous measurements in various regions of the radiation belts. The launching of the satellites *Electron 1* and *Electron 2* is the first step in this direction.

A clarification of the nature of the "near-Earth cosmic accelerator" will make it possible to solve most important scientific problems. It is already known that there exists a "cosmic accelerator" of incomparably great magnitude. During the so-called solar flare-ups, the Sun is influenced by a "cosmic accelerator," the force of which exceeds a thousand times the force of the "near-Earth accelerator." This solar accelerator produces particles with energies up to 10 billion electron volts. In the depth of our galaxy, there exists an "accelerator" which is a billion times larger in dimension and which creates particles with energies up to 10^6 billion electron volts. Finally, beyond the realms of our galaxy there exist "accelerators" which create particles with even greater energies.

In order to understand the chain of these fascinating problems on the creation of high energy particles, which are component parts of cosmic rays, it is necessary to begin with the most accessible region of near-Earth cosmic space.

The selection of orbits for the space stations *Electron 1* and *Electron 2* was based on the necessity of simultaneous investigation of the upper layers of the atmosphere, the Earth's radiation belts, and near-Earth cosmic space.

A number of other factors were thereby taken into

consideration, as for instance the conditions of radio contact during the period when the spacecraft transmitted information to Earth, the duration of the spacecraft's stay in orbit and the spacecraft's exposure to the Sun.

As a result of the consideration of a number of possible variations for the Electron space system, two elliptical orbits with great eccentricity were selected. Such orbits secure the execution of scientific investigations across a range of altitudes from the upper layers of the atmosphere to cosmic space beyond the limits of the radiation belts. Such a range is of extreme interest. The first orbit is located within the most interesting regions of the internal radiation belt. This orbit partially crosses the external belt and takes in the area of space which has an irregular magnetic field, i.e., where unstable fluxes of particles which cause polar luminosity are formed. The second orbit partially passes through the internal belt, the most interesting regions of the external radiation belt, and then follows through regions located beyond the external belt and which have nonstationary fluxes of low energy electrons which are known in literature as the most extreme belt of charged particles. The apogee of the first orbit was selected to be 7000 kilometers which approximately corresponds to the external limits of the internal radiation belt. The apogee of the second orbit was selected to be within the limits of 65,000 to 70,000 kilometers. The perigees for both orbits were established within the range of 400 to 460 kilometers.

It is essential that the focal axis of the orbits of the space stations [i.e., the lines which connect the perigees with the apogees] be selected in varying directions. In the case of a low orbit, the location of the focal axis corresponds to the condition of a most convenient situation in regard to the internal radiation belt. In the case of a high orbit, the position of the focal axis within the orbital plane is selected so as to obtain the highest possible variation of altitudes at identical geographical latitudes during the flight on the ascending and descending orbital trajectories, which is important from the viewpoint of scientific measurements during the investigation of the external radiation belt. The inclination angle of both orbits is about 61° toward the equatorial plane. The magnitude of inclination has a strong influence on the orbital parameter changes under the effect of lunar and solar disturbances and also due to the oblateness of the Earth. In the case of the selected inclination, the orbital perigee will shift toward the north in time and what is more important, the orbit of *Electron 1* will pass through the entire thickness of the internal radiation belt during such a shift of the focal axis.

The location of the orbital perigees in the northern hemisphere assures the most favorable condition for conducting radio contact between the space stations and tracking points on Earth. At the same time, when the spacecraft is located in the area of the orbital perigee, the volume of scientific information is at its maximum since in this area measurements related to the study of the upper atmosphere are conducted in addition to the investigations of the radiation belts.

It is known that artificial Earth satellites which travel along low orbits have a limited life span due to the braking effect of the upper atmosphere layers. With the increase of the orbital altitude, the satellites' life spans increase since the braking effects of the atmosphere are decreased. In the case of altitudes with a perigee corresponding to the orbits of the Electron system, the braking effect on the part of the atmosphere can be disregarded. However, by increasing the altitude of the orbital apogee to several tens of thousands of kilometers, new factors begin to exert their influence quite noticeably on a satellite. These factors are the gravitational forces of the moon and sun. Obtained calculations have shown that in the presence of an unprovable combination of such forces, the life span of a satellite on an orbit with an apogee of 65,000 to 70,000 kilometers is several days.

To continue this crucial scientific investigation, Soviet scientists on July 11, 1964, sent the highly instrumented *Electron 3* and *Electron 4*—each weighing about a ton—into separate orbits from the same booster. *Electron 3* circled between 251 and 4,365 miles above earth, and *Electron 4* between 379 and 41,078 miles. From a scientific standpoint, as opposed to public-exciting spectaculars, the Electron series was one of the most imaginative and practical experiments yet conducted in space.

Somewhat related to the Electrons were the gargantuan Protons, by far the heaviest satellites ever orbited. The first Proton, weighing as much as an empty Greyhound bus, 26,896 pounds, was hoisted into orbit on July 16, 1965. This was followed four months later by the 12-ton *Proton 2*, launched on November 2, and by *Proton 3*, whose weight was not announced, launched July 6, 1966. U.S. observers were initially more impressed by the size of the new Proton boosters (estimated at 3,000,000 pounds thrust) than by their formidable scientific task. The Protons were, in essence, massive orbiting physical laboratories whose special purpose was to investigate cosmic rays of very high energy. The Soviets were searching for energies ranging up to 100 trillion electron volts, or more than 1,000 times the energy of the largest accelerator now planned on earth, the 200 billion electron volt machine. Soviet scientists also say that *Proton 3*, especially, is searching for quarks, subnuclear particles never seen or proved to exist but believed theoretically to exist. As researchers in the field of cosmic rays and as hunters of quarks, the mammoth Protons could have scientific implications of far more importance than the manned flights that have commanded the headlines. A number of the world's most prominent nuclear physicists maintain that civilization is on the verge of unlocking the secrets of superenergetic particles and thus giving man, for the first time, a basic understanding of the fundamental nature of matter.

Much of the weight of the Protons comes from the shielding required. *Proton 1* apparently consisted largely of a heavily shielded detection apparatus, or ionization calorimeter, that sorts and measures cosmic particles. *Proton 2*, as a high-energy physics laboratory, contained one 10-foot by 6-foot instrument bank capable of measuring the energy spectrum up to 100 trillion electron volts and determing cosmic-ray chemical composition. Details of *Proton 3* are unknown except that instrumentation was designed to isolate quarks.

The main body of the Proton was a hermetically sealed cylinder with convex front and back. Four solar panels extended like giant tennis rackets from the forward end. Inside were an ionization calorimeter, a spectrometer to measure moderate-energy cosmic rays, and a special instrument to record high-energy electrons. Experiments and spacecraft operations were performed both automatically and by ground command. Attitude control was maintained by lowrate spin-stabilization using a variety of orientation sensors and gas jets. Protons used the proved Soviet heat exchanger and insulation techniques.

The most important scientific findings, the Soviets subsequently reported, were that measurements were taken of nuclei of new particles and that cosmic rays were 100 million years old. It is assumed that had Soviet scientists made any other startling, nonmilitary discoveries with the Protons (and a number of smaller Cosmos satellites that appeared to have similar purposes) they would have announced the fact by now at one of the international symposiums in which they participate with increasing numbers and frequency.

In the general field of high-energy particles, U.S. and U.S.S.R. scientists generally enjoy an unusually close professional relationship. Such U.S. scientist-administrators as the AEC's Dr. Glenn Seaborg, Harvard's Dr. Norman Ramsey, and Berkeley's Dr. Albert Rosenfeld have lectured at several of Russia's nuclear centers, and the U.S.S.R. has sent to similar U.S. facilities such outstanding nuclear physicists as Dr.

Gersh Budker of Akademgorodok in Siberia. Budker formerly worked at the nuclear-energy center at Dubna, about 85 miles north of Moscow. When I visited Dubna in 1966, I saw evidence of the keenness with which Russian physicists regard their professional competition with their U.S. colleagues. In the second-story office of physicist Vitily Karnavkhov, there is a sort of chart or box score of recent major breakthroughs in nuclear physics. The Dubna discoveries on this East-West scorecard are written in red; Western discoveries are written in blue. The partial and disputed Dubna discovery of the 104th element, they will tell you, won the seventh inning for the Soviet Union, and they are delighted when American physicists argue back with them.

The inauguration of the Soviet communications-satellite program furnished another illustration of the way the ambiguous "Cosmos satellite" label works in their favor. In August of 1964, Russia launched a satellite they called *Cosmos 41*—actually a communications satellite that failed. But by April 23, 1965, there was another try, and this time they were very successful. *Molniya 1* (*Lightning 1*) appeared after the U.S. Telstar, but unlike comsat, the maneuverable Molniya was equipped to relay film and color television, as well as to take high-altitude pictures of earth. After a little over a week, Molniya was observed to change its orbit, like Polyet, to a new apogee of 24,773 miles and a perigee of 339 miles. The maneuver not only kept its orbit in the Northern Hemisphere, but also increased the orbital period to exactly a half of a day (from 11 hours 48 minutes to exactly 12 hours). This contrasted with the demonstrated U.S. method of maintaining Syncom over a fixed point on the equator.

The Soviets had carefully prepared for the success of Molniya; in fact, they may even have rehearsed it with Cosmos satellites. On its first orbit, Molniya relayed a filmed television broadcast from Russia's warmest port, Vladivostok, to Moscow—a distance of more than 4,000 miles. Later, Molniya relayed live color television over the same distance. I reviewed the Molniya film and tape in Russia, and picture and color quality were remarkably good. These latter were compatible with the SECAM system that is being developed jointly by Russia and France in competition with another system proposed for Europe developed by the U.S. and West Germany. The Russians were late in the communications-satellite field (unless they had secretly developed a Cosmos-series military system), but the color proficiency of *Molniya 1* quickly established them as highly competitive in the field of electronic communications.

A little-publicized experiment of Molniya was auxiliary to its primary purpose as an active-repeater comsat to relay television, voice, facsimile, and Teletype transmissions. Molniya also had exterior motion-picture, television, and still cameras with interchangeable lenses operating on command from earth. The cameras were aimed at earth by a special autonomous system that functioned independently of the satellite's earth orientation system. One lens is reported to have covered the entire visible portion of the Northern Hemisphere. On May 18, one of the cameras sent to earth the first television pictures of our planet—a cloud-fuzzed view of the Atlantic and Pacific oceans as seen from a great height above the North Pole. Viewers could see the sun shining brightly off the Arctic Ocean.

Several other Soviet earth satellites had particular functions worth noting. *Cosmos 5*, for instance, was able to report on the decay of the high-altitude radiation created by the U.S.S. Starfish nuclear test over the central Pacific on July 9, 1962. The Soviet Academy of Sciences indicated this new purpose of the Cosmos series by announcing "one of the basic missions of Cosmos satellites is to study and control radiation hazards to manned flights, especially after high-altitude explosions." The academy claimed *Cosmos 5* "obtained a detailed picture of the results of the American atomic explosion of July 9, 1962." Fortunately for both ordnance specialists and satellite policemen, the treaty outlawing atmospheric nuclear tests was signed on October 10, 1963.

Typical of the brevity and vagueness of the Soviet announcements of many of the Cosmos satellites was the launch of *Cosmos 100*, which Tass reported was "equipped with scientific instruments to continue space studies." However,

Cosmos satellites place themselves in categories of sorts by the inclinations of their launch and the place from which they are launched. KY-49s are those Cosmos satellites launched from Kapustin Yar at 49° (or other than 65°) to the equator, and having relatively long lifetimes. They are believed to weigh between 1,000 and 2,000 pounds and to have carried the main load of early Soviet atmospheric research. However, the recent absence of scientific reference to them suggests their purpose has become more sensitive. TY-65s and 51s are launched from two sites near Tyuratam, have payloads of over 10,000 pounds, and are believed to be Vostok-type reconnaissance satellites. They could be unmanned or manned by muted pilots. Most are put in low orbits, and their payloads are known to re-enter after just a few days (usually eight days) before their booster does. The longest duration member of this group was *Cosmos 20*, which orbited for twelve days. Other eccentrics include *Cosmos 41*, which had an apogee of 24,765 miles and may have conducted synchronous corridor tests; *Cosmos 50* and *57*, both of which exploded (or were purposely shot down) in orbit. *Cosmos 57* exploded into more than 150 observable pieces. Also eccentric were those, such as *Cosmos 111*, that were suspected of having been abortive lunar probes.

Considered eccentric also are the triple-payload Cosmos satellites 38–40 and 54–56 and the quintuple payloads that began with *Cosmos 72*. Of these quintuple payloads, *Cosmos 81–84* are particularly mysterious because it is not known what purpose their electric generators powered by a radioactive isotope served. The one-day flight of *Cosmos 47* was believed to be an unmanned trial run of the Voskhod flight that came a week later. Practically all Cosmos satellites launched after January 1, 1966, were in polar orbits, ranging from 88° inclination to 99° to provide maximum earth coverage.

On February 22, 1966, the Soviets broke through the Cosmos-series security somewhat with some facts about *Cosmos 110*. This satellite, launched from Tyuratam, orbited two dogs from 116 miles to 562 miles high for 330 revolutions. The dogs, Veterok (meaning "breeze") and Ugolyok (meaning "little bit of coal"), had electrodes grafted to their peripheral sinus nerves, and probes inserted into their arteries to measure and to serve as a needle for the introduction of pharmacological agents (believed to be a new antiradiation serum administered for comparative purposes to Veterok). Other more standard devices measured respiration and heart action. Since the dogs were purposefully orbited very high into the radiation belts, on-board dosimeters kept track of their exposure rates. They were fed by stomach tubes and could lie, sit, or stand.

When Veterok and Ugolyok were returned after 22 days of weightlessness and high-count radiation exposure, they showed more serious reactions than any man or dog had before them. Their muscles had weakened; they were dehydrated, had lost calcium, and had difficulty walking. It was five days before they returned to a condition approaching normal and ten days before full recovery.

Russian doctors, including Boris Yegorov, were cautious in their statements, but the results seemed to confirm some long-expressed fears of Soviet scientists about dangers of prolonged weightlessness and elevated radiation exposure. "These effects," one of them said, "may not apply to man, but they raise new problems for space medicine."

About a month after Veterok and Ugolyok landed, science writer and editor Bill Beller and I questioned the man chiefly responsible for their flight, Academician Vasily Parin. We asked him to explain Soviet policy in rather suddenly reverting to the study of animals in space. Parin indicated that, far from transferring the majority of their biomedical studies to man, they intended to make extensive use of animals. He suggested the chief reason for this was that their studies were now designed to determine the types and rates of adaptability of organisms to space. "Is it," he asked, "like stepping into a cold room? Or is it like walking slowly into a colder climate? We will use animals to find out." His reaction to the *Cosmos 110* experiment, plus the reaction I received the next day from Dr. Oleg Gazenko at the Soviet Academy of Sciences, led me to believe that the Soviets are seriously perturbed about man's ability to function effectively after extended weightless flight, especially if radiation levels are elevated. "There is a combination of factors," said Gazenko, "that is difficult for the body to adjust to."

Just as the Soviets used their unmanned satellites to acquire additional data on what, to them, was the disturbing and complex effect of the space environment on living organisms, so also did they use them to test highly sophisticated engineering concepts that would apply to both the planned manned and the unmanned flights of the future.

One such famous flight provided a partial answer to what had been for years the most tantalizing enigma of the Soviet space program—their long-postponed attempt at full rendezvous and docking. The partial answer came in a surprising fashion.

By October of 1967, the tenth anniversary of *Sputnik 1*, there was neither verified nor unverified evidence that any two Soviet satellites had ever come closer than the three miles obtained on the Bykovsky-Tereshkova flights of 1963. Yet Russian scientific literature continued to support heavily both the construction and the resupply of large space stations and their manned landing on the moon. The persistent question was: How could either be accomplished without rendezvous and docking?

During the tragic flight of *Soyus 1* on April 24, 1967, those Moscow correspondents, particularly those of UPI, who had painstakingly developed sources with some record of reliability, had been confident that Komarov was to have been joined in space by a second manned Soyus. Unforeseen malfunctions, in other words, aborted the first scheduled mating in space of two Russian vehicles. Although, as we have seen, the Soviets seldom repeated experiments, few, if any, Western observers expected the first rendezvous and docking to occur with other than manned spacecraft. The very presence of a man aboard was widely considered the necessary insurance for success of this delicate maneuver. But, once again, the Russians confounded expectation.

On October 27, just eleven days before the elaborate ceremonies marking the fiftieth anniversary of the Russian Revolution, a windowless spacecraft shaped like an elongated bell was orbited from Tyuratam. It so precisely followed the *Soyus 1* inclination and altitude that some U.S. observers suspected that this was a redesigned Soyus which could be manned and which might repeat Komarov's suspected goal of rendezvous and docking. But radio monitors reported no voice transmission from space. Three days later, on October 30, a second spacecraft appeared on radar just 15 miles from the first. The second spacecraft, the female of the pair, resembled the first externally, except that its coupling cone at one end was larger in order to receive the stamen-like projection of its predecessor. Each contained aerials at its fore end and a pair of gull-winged solar panels that unfolded from near its bell-shaped base. True to one of the legends of man, it was the first satellite aloft, the computerized male, which now pursued the female around the earth. Mating took place in the female's first orbit over the South Atlantic. The Soviets announced that the male, *Cosmos 186*, had oriented its electronic stamen toward *Cosmos 188*, and inserted it for a rigid union that linked electrical systems. Russian ground controllers watched the linkup through television cameras mounted on *Cosmos 186*.

The coupled pair carried out commands from Russia for three and a half hours. Then, on another command, the two spacecraft separated. Each reignited its engine and entered a new orbit. Twenty-four hours after the linkup, the male returned to a soft landing on earth, presumably using parachutes and the same obscurely labeled "ionic velocity vector formers" that had braked the Voskhods to a gentle landing. Three days later, the female returned to earth by the same means.

The Soviets supplied few details, and, as this goes to press, it is too early to try to put the pieces together from scattered fragments of their scientific literature. The fully automated, remotely controlled coupling was patently more complicated than the manned rendezvous of the U.S. Geminis. Both Soviet and Western reaction suggested the obvious and highly significant application of the automatic docking to the assembly of space stations and of lunar and planetary launching platforms. Not mentioned was another application of possibly equal significance. Presumably the pursued female carried beacons and possibly radar and other devices ancillary to the union. But the male, *Cosmos 186,* may also have been the prototype automatic interceptor spacecraft. With reasonable modifications, such a homing vehicle may soon be able to demonstrate its ability to intercept, inspect, immobilize, destroy, or capture any object in space. If we add to the enormous launching variables of launch place, time, and inclination, an orbiting computer which can maneuver into the plane of an inert orbiting target and close on its electronic image at a desired rate, we have a predator of space not too unlike a night-flying interceptor aircraft. The uses of such a homing orbital predator are manifold, not the least of which—in the tradition of the destruction of the high-flying U.S. U-2 reconnaissance aircraft—is the interception of reconnaissance spacecraft. From reconnaissance spacecraft to manned spacecraft to orbital bombs to ballistic ICBMs to fractional orbital bombs represents a feasible progression. And it is a progression that will assuredly make obsolete the bulk of existing defenses against trans-space military payloads.

Thus the achievements of *Cosmos 186,* both scientific and military—which are often the same—must take their place among the major milestones of the first decade. The Soviets not only did not repeat any of their previous experiments, they transcended, rather than repeated, our own experiments.

The U.S. reaction was again surprise and an attempt in several quarters at familiar and mollifying rationalization. According to the Associated Press, Dr. Edward C. Welsh, principal space adviser to the White House, said that the U.S. had the technical ability for a number of years to achieve an unmanned docking but saw no need to do it. While Dr. Welsh may not have intended this as the pale palliative it sounds like, it is one more example of those well-intentioned U.S. spokesmen who appear to react to Soviet space achievements with the partisan thinking of public-relations men.

Venus 3 on exhibit at the Cosmos Pavilion of the Economics Achievements Exhibit in Moscow. A similar-spacecraft with its spherical landing vehicle three feet in diameter, dropped by parachutes, deposited the pendant with the Soviet coat of arms on the surface of Venus in March, 1966.

Andrian Nikolayev, pilot of *Vostok 3*, and his wife Valentina Tereshkova, pilot of *Vostok 6*, with their daughter Alenushka. The romance and marriage of the two cosmonauts was given wide coverage in the Soviet press.

14
To the Moon and Planets

THE Russian determination to explore the moon and inner planets, as indicated in Chapter 6, is deeply rooted in its national consciousness. The popular and scientific adoption of the romantic notion that man has a significant destiny away from his home planet is explicit in the pre-Sputnik Soviet film *Blazing a Trail to the Stars*, in the public statements of Sergei Korolev while under his anonymous cloak of chief designer, and it is even more explicit among those Soviet poets, philosophers, artists, writers, and journalists who adopted and embraced their new manifest destiny.

Some of the early Soviet journalistic fascination with the subject of a rocket flight to the moon was loftily dismissed by American experts as Sunday-supplement stuff when much of it was actually on solid ground, especially material by Yu. S. Khlebtsevich, the author of *The Road to the Cosmos*. V. A. Egorov's announcement that from 1953 to 1955 Soviet computers had calculated more than 600 trajectories to the moon was dismissed as untrue or, if true, a fanciful intellectual exercise.

Russian preoccupation with the moon was many-faceted. The military, of course, were interested, but they did not dominate the progression that caused the next rocket after *Sputnik 3* to aim for the moon—then and now a questionable "high ground," militarily speaking. Nor were the curiosities of scientists nor the air pollutions of propagandists the primary motivation. The primary and most powerful motivation

H   213

is rooted somewhere deep in the psyche of a society that is powerfully drawn to the unknown cosmos that lies beyond the heavens. When Gherman Titov parried a question with a shrug by saying, "I didn't find God out there, if that's what you mean," no one in Russia was surprised or hurt. His mission, in their terms, was not to find God but to help them identify with the forces and unexplored bodies of creation. And this identity could be realized at all levels—from the consciousness of a peasant gazing at the moon, to the flag-waving politician, to the artful propagandist, to the absorbed scientist plotting his fascinating new curves of penetration. The Soviet being is able, almost universally, to identify with the apolitical world of celestial bodies within his reach.

The Russian writer, Vladimir Orlov, touches some of these Soviet sensibilities and susceptibilities in his self-appointed role as the Soviet philosopher of space. Orlov requested and got early permission to observe the architects of the space age in action. His chronicles are unfortunately tinged and marred by his preoccupation with international one-upmanship. But Orlov does reflect this peculiar Soviet passion. "These celestial bodies created by the genius of man," he writes, "are growing in number. . . . These man-created planets are being populated by the simplest organisms, plants and animals. Man has appeared on artificial planets. A majestic act of creation is underway—an act which the Bible has ascribed to the unlimited omnipotence of God. Even the Pope of Rome is in a hurry to get into the same boat with the miracle workers by interpreting the conquest of the cosmos as a supreme feat of Deity. It is hard to understand though why the Lord is so eager to help the atheists! What does take place is an unheard-of expansion of that form of highly organized matter that is called 'Life.' There was a time when terrestrial life stepped over the threshold of the ocean and conquered the land; now it has stepped off the Earth to conquer the abyss of the cosmos."

According to Orlov, Soviet moon and planetary probes are "a younger sister of the Earth and a daughter of the Sun." In his prostration before the throne of space, he points out that "the total of all motors in the Gagarin ship totaled 20 million horsepower. This means that the chariot of the Soviet cosmonaut was drawn by all the horses of tsarist Russia at the turn of the century." Orlov also exorbitantly praises the space-emancipation of the Soviet worker. "Also incarnate in the rocket," he reports, "is the new image of the worker in the cosmic era, where the boundary between the physical and intellectual effort is becoming less conspicuous; when a mechanic becomes sculptor; a fitter turns into a surgeon; and a cutter becomes a mathematician controlling an electro-programmed lathe. Who is so bold as to tell apart the physical and the intellectual when a worker cherishes a machine detail as Stradivarius cherishes his violin? —or when a shop worker informally confers with a scientist while the light of common inspiration shines on their intent faces?" In aiming for the moon and planets, the Russian worker, according to Orlov, has adopted the following solemn and dedicated motto: "To Work Without Spoilage, Because of the Urgency."

To the scientists and engineers who attempt to convert the Soviet manifest destiny into functioning and reliable hardware has fallen the task of matching the cosmic reach of the imagination to the mundane grasp of engineering. In the case of lunar and planetary probes, this has proved a difficult task indeed. In no area of space exploration have the Soviets had more failures, yet in no area have they exhibited more dogged determination. Their very boldness invited both failure and success. As reported in an earlier chapter, *Luna 1*, which followed on the heels of the early Sputniks, by side-swiping the moon and orbiting the sun, set the bold pace. In that same year, 1959, *Luna 2* hit the moon and *Luna 3* photographed its dark side.

Soviet writer, K. Leonidov, in describing the atmosphere surrounding the early Soviet deep space probes, seems, at times, almost to be chronicling the unconscious formation of the elite cadre of an extraterrestrial society. He reports, in addition to human fallibility, a sense of urgency and a sense of mission that is almost transcendent. On one occasion, he reports, after mechanics and rocket technicians had retired to the *perekur* (smoking lounge) to await the imminent rocket launch, a red light began to blink wildly at the control panel: "Defective! Defective! Defective!"

"From the dumb cry of the light," Leonidov writes, "nerves become strained like a violin string. The heart begins to pound violently. A taut silence advances on the command post. The job could not be simply to find and eliminate the defect. It was important to fire the rocket at precisely the appointed second and there remained only one minute until launching. It was necessary to decide quickly. Turning pale at the defect, the commander of the launching team rushed to the chief designer. And the following order was heard at once: 'The rocket launching will be suspended. We will determine what has occurred. Report immediately.'

"The radio began to thunder on the rocket field summoning the engineers and mechanics from their sheds. Like doctors examining a gravely ill man, they felt around all of the corners of the rocket. The people who had launched the first moon rocket stood quietly now at the foot of their new rocket. How many sleepless nights had they spent on the experiment striving for a smooth and reliable assembly!

"After a little while the reasons for the failure became clear. The voice of the chief designer was heard. 'Dejection cannot be weighed and heart is not lost. The rocket will be taken away and another of the same design will be brought out. We will examine it carefully and arrange a double control.'

"What was understood by this?" Leonidov continues. "With new strength and with doubled energy work proceeds furiously. A new rocket was set up at the edge of the forest. The engineers and mechanics covered the spot like ants. . . . The hatches of the rocket were battened down. Everyone went for cover and was glued to the loud speaker. The gun-layer, Aleksei, checked his work. When they heard the command for launching, everyone stiffened. Suddenly a voice was heard, 'Contact!' And at the same instant the deafening roar of the motors was heard. The launch had been achieved within only one second of the assigned moment."

Later, as rocketmen gather around a loud-speaker, Leonidov quotes the breathless announcement of impact: "Comrades, an important event is taking place at this moment. The rocket is swiftly rushing to the moon. In one or two minutes the moon will take into its embrace our Soviet rocket. One minute is left. . . . Comrades, do you hear me? It has just been made known. At zero hour, two minutes and 24 seconds, Moscow time, the second Soviet rocket has reached the surface of the moon. . . . Do you understand? Do you hear me? Do you understand? What are you doing? I congratulate and embrace all of you."

Such breathless exuberance, bordering on ecstasy, translates with difficulty to Western ears, especially to those Westerners who have not participated in the launch of a rocket toward the moon. Yet Leonidov's prose description comes close to depicting the almost childlike rapture with which Russians respond to a lunar probe, even when there is no heartbeat and respiratory rate emanating from space. The rocket and its cargo is obviously highly personified. And as the Russians played the song "My Motherland Is a Great Country" after this particular launch, they are unafraid of emotion in the presence of unmanned astronautical events. They were quick to construct around such occasions a ritual, a tradition, and appropriate symbols, culminating in Moscow's graceful and splendid titanium monument to space. Often at Cape Kennedy, we developed and used symbols also. Some of our rockets bore, in addition to dozens of signatures, Saint Christopher and other religious medals, and such slogans as "Love Lifted Me" and "Have Ball; Will Orbit." The Soviet lunar rockets also bore innumerable signatures and slogans, as well as historical mementos they expected eventually to recover from the surface of the moon.

After their spectacular early success with three lunar probes and after they began to experience failures in their unannounced attempts to reach Mars and Venus, the Soviets did two things characteristic of their passionate involvement with deep space. First of all, they secretly decided that for the next several years they would attempt to reach Mars or Venus on every single occasion each of these planets moved in a favorable position for intercept from earth. They would fire, in other words, through every window that offered them a favorable launch date. Second, to avoid a repetition of speculative failures, they set up a broad designation of a family of space shots they chose to call their "Cosmos satellites" or "the Cosmos series." This gave them the opportunity, if they wished, to mingle planetary experiments—or any other experiments they wished to keep secret—in with the general class of Cosmos satellites largely designed to explore our electromagnetic atmosphere. Since their preferred technique for launching planetary probes was first to put them in a stabilized earth orbit, this gave the Russians the opportunity to avoid mentioning those of its planetary probes that failed to leave earth orbit or that failed to leave it at the necessary speed or trajectory. This policy has made tabulation of Soviet deep space probes difficult, although the Russians over the years have gradually and consistently become more frank —both in acknowledging their own failures and in reporting their findings.

Those Americans who attempted to keep the Soviet score estimated that fully one-quarter of all Soviet rocket launches in the first eight years of space exploration were designed for deep space penetration (i.e., to thrust beyond the most common orbital belts of earth satellites). It is certain that they scheduled more deep space probes than we did, partly because we entered into a more time-consuming and perhaps more thorough testing of the ability of vital components to stand the long-duration bake-freeze of deep space probes. The key components that early and repeatedly failed the Soviets were in communications. They were also penalized by the inability of extremely remote operational equipment to function after more than 48 hours.

Except for their first three Lunas, the Soviets' overall record up until the spring of 1966 was rather dismal. On February 12, 1961, they put the 1,419-pound *Venus 1* probe— packed with miniaturized cosmic-ray, magnetic-field, charged-particle, and solar-radiation instruments—into an earth-parking orbit. Then they fired *Venus 1* from its earth orbit on a scheduled multimillion-mile journey to the evening star. But Russian scientists lost all contact with their ambitious probe when it was only 4,700,000 miles out. Unknown to the U.S. public, the puzzled Russians went to the unusual length of sending a disheartened delegation to England's radio observatory at Jodrell Bank. They wanted to see if they could detect signals from their lost probe as it sped in the general direction of Venus. But they were greeted by nothing but silence. In exchange for this unusual privilege, they later invited Jodrell Bank's director, Sir Bernard Lovell, to visit several Russian tracking stations, installations he later made a point of claiming to be inferior to those of the West.

Russia claimed a partial success from the muted *Venus 1* when they later estimated it passed within 62,000 miles of the moist planet.

On November 1, 1962, during a favorable Mars window, the Russians, in a delayed announcement, confirmed the launching toward the red planet of an ambitious instrumented and photographic probe weighing nearly a ton. Called *Mars 1*, the pioneering probe was launched from a parking orbit. The use of a parking orbit, M. V. Keldysh, president of the Soviet Academy of Sciences, once explained, "has opened up new possibilities for interplanetary flights since it eliminates the necessity of choosing specific dates for flights to the moon; it makes possible the launching of heavier vehicles toward Venus and the other planets; and it removes the restrictions connected with the fact that not all points on the earth are equally advantageous for launching."

The 48-million-mile journey of *Mars 1,* which began just after Khrushchev withdrew his missiles from Cuba, was planned for seven months. Its 1,966 pounds were contained in a cylindrical, double-compartmented probe about 13 feet long and about 4 feet wide. It started out close to the prescribed course at the same time the U.S. *Mariner 2* was already speeding toward Venus. The Mars launch was made amid various claims by U.S. observers that, in unannounced attempts, previous Soviet Venus shots had failed on three occasions. NASA administrator James E. Webb had announced the previous September that the Russians had also failed on two attempts during the window of October of 1960 to send probes to Mars. These reports were denied by a spokesman for the Soviet Academy of Sciences. But there was no denying the existence of *Mars 1*. The day after launch, Andrei Severoy, director of the Crimean Astrophysical Observatory, took six photographs of *Mars 1* as it sped away from earth at 2.4 miles a second. The probe and its finalstage rocket, photographed against the night sky, appeared as a 14th- and 13th-magnitude star. By its fourth day, it was more than half a million miles out and continued to radio back data. Aboard were a television device for photographing the surface, a "spectro reflector" to examine the possibility of organic deposits that Soviet scientists largely believed to exist on Mars, and a spectrograph to analyze ozone absorption in Mars's atmosphere. During its space journey, *Mars 1* had equipment for measuring magnetic fields and radiation bands, and a radiotelescope aboard to register streams of low-energy protons and electrons.

During the first month, *Mars 1* was successfully interrogated 37 times and sent back significant data, especially in the measurement of charged particles, to a distance of 5 million miles. *Mars 1* discovered that since the 1959 Soviet flights background radiation had increased in the neighborhood of 60 per cent. On November 30, it encountered an intensive stream of solar particles that reached a level of 600 million particles per sq. cm. per second. It also encountered a severe meteor shower at altitudes around 15,000 miles from earth and registered solar wind and magnetism in deep space. Then on March 21, more than four months after launch and within three months of intercept, all radio contact was inexplicably lost. The only theories the Soviets had were that either its solar batteries had failed or the probe had reoriented itself so that antennas were no longer pointing toward earth. Thus *Mars 1* lost its earth lock at 65.9 million miles, amid widespread speculation that the Soviets had failed in the earlier attempts to command a midcourse correction. It is a safe assumption that its original course, which would have carried it within 162,000 miles of Mars, was supposed to be altered by radio command from earth to bring it within several thousand miles of its mysterious target.

With the beginning of 1963, Russia again launched a determined effort to reach the moon—this time with payloads of about a ton and a half that were intended to accomplish the then extraordinarily difficult feat of a soft landing on the moon. The attempt ran into an unaccountable variety of failures—some in areas the Soviets considered they had previously mastered. For instance, of about eight attempts, 50 per cent failed in the early phases of flight. They either were unable to get into the desired parking orbit or, on one occasion, could not successfully leave the achieved parking orbit. Of the four partly successful ones, two missed the moon entirely—*Luna 4*, launched in April of 1963, by about 5,000 miles and *Luna 6*, launched in June of 1965, by more than 100,000 miles. *Luna 5* crashed on the moon in May of 1965 when its braking retros failed to fire.

One success that was to remind the Russians of their lunar year of 1959 was their deep space probe, *Zond 3*. During a flyby on July 20, 1965, *Zond 3*'s cameras photographed that small segment of the moon's dark side that *Luna 3* had missed in 1959. The Russians could thus claim—in one of their very few repeated experiments—that they had finally obtained photos of the entire side of the moon that is always turned away from the earth.

Later that year, the Russians were sure they would have a success with *Luna 7* that would help fill in the mysterious gap following upon the heels of the Belyayev-Leonov flight in *Voskhod 2*. They scheduled *Luna 7* on an auspicious date, October 4, the eighth anniversary of the flight of *Sputnik 1*. *Luna 7* successfully went into parking orbit, and its 3,318-pound payload was fired out of orbit on a precise moon-intercept course. Some eighty hours later, Jodrell Bank detected the reduced speed close above the moon indicating that retros had indeed fired. But *Luna 7* hit its target so hard that its electronic brains were destroyed.

Russia's difficulties with deep space probes partly carried over to two Venus attempts in 1965. On November 12, *Venus 2* was launched, but it did not accomplish its principal purpose of sending back photographs of the sunlit side of Venus. *Venus 3*, launched three days later, underwent a successful midcourse correction and three and a half months later made man's first contact with a neighboring planet by hitting

Venus. But Russian scientists admitted *Venus 3* did not achieve its essential purpose. It was to have ejected a sphere containing instruments to measure Venus' surface temperature and atmospheric pressure. Hitting Venus was an astonishing feat of navigation, but information on the latter two factors would have settled scientific speculation that has raged for centuries.

Virtually the same disaster that afflicted *Luna 7* occurred on Pearl Harbor Day of 1965, when *Luna 8*'s retros were again observed to fire. But its radios mysteriously went silent minutes later on impact. Russian scientists had been so confident in *Luna 8* that they had revealed their optimism in advance to a few Western journalists. When the radio went dead, they retired to inaccessibility and silence.

It was an often frustrated and grimly determined group of Russians who again groomed a huge lunar rocket for flight in cold late January of 1966. During their long frustrations, they had seen the U.S. Ranger close-up pictures of the moon's cratered surface in both 1964 and 1965. Neither nation had yet soft-landed instruments on the moon, but the Russians well knew that the long-delayed U.S. Surveyor moon probe was close to its baptism in space and its attempt to settle its spidery frame gently on the virginal surface of the moon.

The era of on-the-surface lunar investigation began on the last day of January. After liftoff, the Soviet lunar rocket arced southeast over the southern Russian mountains toward India. *Luna 9* dropped its massive first stage and was inserted into the desired earth orbit. Then about an hour later, at roughly the two-thirds point of its initial orbit, a powerful burn, augmented by its already considerable orbital velocity, propelled it on a moon-intercept course. The moon-bound vehicle, which weighed about the same as a small truck—3,500 pounds—contained a delicate 220-pound automatic station in a dome atop the propulsion and deceleration section. Detachable pods contained the automatic piloting units designed to control the strange craft until it neared the moon; several micro engines were connected for in-flight changes and orientation.

It was soon apparent to observers in the underground Russian control center that the present course of *Luna 9* would cause it to impact on the moon but not at the desired point. The Russians wanted the impact to occur, for photographic purposes, just in advance of the dawn of the two-week lunar day. For communication purposes, they also wanted the planned soft landing to occur at the highest possible point on the horizon in relation to its earth-based control center, which would send commands to *Luna 9* if it survived the landing. Their preferred target, therefore, was a crater-pitted equatorial mare called the Ocean of Storms, located to the moon's far left as seen from the earth. From previous experience, the Russian scientists knew—before they fired the course-correcting engines—that a speed change of just three inches per second would alter the lunar impact point about eight miles. The length, direction, and power of the burn was highly critical if they were to hit their target in the Ocean of Storms. Near the earth midnight of the second day out, mathematicians rechecked their figures and announced they were ready. But they had to be sure *Luna 9* was in the proper position—or attitude—for the crucial burn. They sent instructions by radio that *Luna 9* obediently followed. First, by using a system of light-seeking lenses coupled to its micro-engines, *Luna 9* oriented itself in relation to the sun. Then a second and more sensitive optical system searched the sky for the light of the moon, found it, locked on it, and expertly aligned the spaceship so that, in effect, it was traveling sideways toward the target moon. With *Luna 9* stabilized, another radio command ignited the course-correcting engine. When the preset correcting speed was reached, *Luna 9*'s control system automatically switched off the engine. When new trajectory calculations were completed, *Luna 9* was found to be right on course. Even though the Ocean of Storms was still more than 150,000 miles away, the spaceship's navigation was completed.

But there was no chance for the Russians to relax, for the task that remained—they had discovered the hard way—was the most difficult of all. The problem that faced *Luna 9*'s human and mechanical controllers on February 3 was brutally simple. Slow it down an instant too soon and it would stall high and, even at one-sixth earth gravity, topple to destruction; slow it down too late and it would crash. It was closing in on the moon at a fantastic speed—nearly 9,000 feet per second—and accelerating under the increasing draw of lunar gravity. Nothing save its rocket engine could slow it; parachutes or drag flaps were useless in the airless void surrounding the serenely sailing but alien and hostile chunk of rock.

Luna 9's landing sequence actually began about an hour out. At this time, the speeding probe crossed a reference point—5,278 miles from the moon—where the direction to the center of the moon coincided with the exact direction necessary for the retro-rocket braking thrust. The on-board radio altimeter signaled when this reference point was reached. *Luna 9* was turned so that the retro-engine was maintained in the straight-down position by a near-surface alignment process that disregarded the moon and used, instead, earth and sun reference points. At precisely 46.6 miles up, the radio altimeter fired the powerful retro-rocket. It took 48 seconds of braking thrust to slow *Luna 9* to near-zero speed. Shortly before touchdown, according to Russian sources, the automatic moon station separated from its braking mechanism and fell nearby—protected by its own shock-absorber material (probably balsa wood). It probably rolled slightly, since it was now as round as a beach ball, finally coming to rest in a shallow crater just to the left of a lunar mare. It had landed, as planned, in the lunar dawn. The slowly rising sun was just 3° above the eastern horizon.

For 4 minutes and 10 seconds it just sat there, as the world of both East and West wondered if it had survived the landing. It was then, as writer Vladimir Orlov described it when he had first inspected it on earth, "like a plant seed visible under a microscope, a seed of the earth. On the moon's surface it is supposed to spring to life under the first rays of the morning sun, spreading out the stamens of its aerials."

Now, as Orlov had described it, *Luna 9* came to life. First, four petal-like segments folded and peeled down until, touching the moon's surface, they forced *Luna 9*'s television scanner into the topmost position. Simultaneously, four antennas, each about four feet long, extended skyward and instantly relayed to earth the fact that the hermetically sealed station had not ruptured on impact, that internal temperature was under control, and that the instruments were functioning properly. Thus, the anxious Russians learned, just before their midnight of Thursday, February 3, 1966, that earthmen had finally achieved a soft landing of instruments on the moon.

Its purpose, of course, was not just to land but also to use its expensively obtained and privileged position to scout the lunar terrain and to inform earth of its findings. Eight hours after landing, as the sun slowly rose, a radio command from earth opened the lens of *Luna 9*'s television camera. It was pointed southeast across a line of long shadows made by the sun. The camera's view—about two feet above the surface—was essentially that of a man, had one been sitting on the lunar surface. Narrow strip mirrors in front of the camera reflected other portions of the surface.

As the automatic station prepared to send the first historic picture back to earth, British scientists and engineers at England's Jodrell Bank completed an ingenious improvisation. They had borrowed a telephoto facsimile machine from a newspaper office, and now they completed the linkage between the machine and the huge and sensitive radio telescope tuned to the frequency previously announced by the Russians—183 megacycles. Soon the signals began to come in, and their facsimile machine traced a clear close-up view of the moon, showing its porous, pitted, crater-strewn surface and small rocks that cast long, sharp shadows. Two conclusions were obvious immediately to both British and Russian observers on earth. The moon was not covered with a thick layer of dust, and since the 220-pound (earth weight) *Luna 9* had not sunk, the surface could support the weight of a man. The Russians had never signed the international copyright agreement. If they had, they could legally have demanded an exceptionally high price for the British release of the *Luna 9* pictures. But the British, bound by no copyright agreement, scooped the Russians on their own pictures by releasing the *Luna 9* photographs first. The series of about nine pictures received by both the British and the Russians showed an odd difference between the last picture sent on Friday and the first one sent on Saturday. The horizon angles did not match. Had *Luna 9* lurched? Did it roll? Or was it jarred by a small quake on the moon? After I had inspected a *Luna 9* replica in Russia, CBS cabled me to interview Dr. Alexandre Lebedinsky, professor of astrophysics at the University of Moscow, on the *Luna 9* accomplishment.

Lebedinsky is a short, intense, and courteous man with reddish-blond hair. Like Dr. Harold Urey in the U.S., he is one of the handful of scientists who first see and evaluate new lunar data. Lebedinsky told CBS that Soviet physicists did not believe there was present volcanic activity on the moon or that the *Luna 9* lurch was caused by anything like a moonquake. Lebedinsky felt that as the sun rose, the increasing temperatures of the moon's surface, plus the station's continued weight, caused a gradual breakdown of the gingerbread-like lunar soil beneath *Luna 9*. Whatever its cause, the lurch was a fortunate occurrence because, Lebedinsky pointed out, the slight shift in camera angle permitted stereoscopic analysis of some of the photographs.

Other *Luna 9* instruments, besides the television camera, were dosimeters and other devices for sampling and measuring radiation and other characteristics of near-moon space. The chief lesson taught by *Luna 9*, according to Soviet scientists with whom I discussed the subject, was that the surface was adequately stable and that the daily radiation dose was an unexpectedly low and harmless amount—about 30 millirad. Soviet scientists also took an unequivocal stand on the long-debated question of lunar water. Water, they now confidently postulated, had once been present in great abundance and may still be there as crystallization water held chemically by rocks. On the question of volcanic activity, most felt that lava from ancient volcanoes had interacted with water, salt, volcanic ash, and meteoric dust to create the porous but solid surface on which *Luna 9* had landed. As Alexandre Lebedinsky proudly showed me the freshly printed strip of nine *Luna 9* photographs, he emphasized over and over that facsimiles were being turned over to U.S. and world scientists for study and analysis. "We have studied your Ranger photographs," he said, "and you should study ours. Scientists should work together."

Approximately two months after the historic soft landing of *Luna 9*, the Russians again sent a scientific payload to the moon, but this time they chose to orbit instead of land.

Launched April 14, the first part of the *Luna 10* flight occurred in much the same way as that described for *Luna 9*. The whole assembly looked somewhat like an old-fashioned pot-bellied stove hurtling through space. The midcourse maneuver, instead of aiming for a specific point on the lunar surface, aimed at a point in space about 200 miles behind the speeding moon. The retro-rocket fired as the 539-pound *Luna 10* approached this point, reducing the speed to about two miles a second. This was just enough to cause *Luna 10* to be captured by lunar gravity and, in fact, to become a circling prisoner of the moon. In becoming the first satellite of a satellite, *Luna 10* had performed an amazing feat. The precision required could be symbolically compared to bringing in an airplane at a given speed and direction on an invisible runway 240,000 miles away. The initial high point (apolune) of the lunar orbit was 621 miles; the low point (perilune) was 217 miles. The Russians considered this an especially desirable orbit for their proposed close-up study of the stability of the moon's gravitational and magnetic fields. They also had instruments aboard to study meteorite activity and the nature and radiation of the moon's surface.

The first raw data sent back by *Luna 10* astounded even the Russians. Its radios reported unexpectedly high cosmic radiation in the space around the moon. The reason for this lay in the time the Russians had chosen for placing *Luna 10* in orbit. It was no accident that at this time the sun, earth, and moon were all in line and in that order. The Russians immediately concluded that the high-radiation count was caused by the fact that the earth's radiation field was not circular but comet shaped and that, from time to time, the moon hitchhiking along with the earth passed into the poisonous veil trailing the earth's dark side. These findings would be of profound importance to any future attempt to set up colonies on the moon. They could even affect the scheduled launching of the first men to the moon.

Luna 10's instruments also firmly established that matter in the lunar surface was basaltic, like that of earth. Gamma rays emitted from the three radioactive elements in the lunar crust—uranium, thorium, and potassium 40—were much like those from the earth's crust. As far as Russian scientists were concerned, this not only established that the moon was once molten (which allowed lighter rocks, such as basalt, to float to the surface), it also meant something of even more significance: the earth and moon, the Russians concluded, have a common origin.

Any good scientific experiment raises more questions than answers, and *Luna 10* was no exception. In fact, it provided the scientists of the world with a tantalizing enigma. During its 2½-day journey from the earth to the moon, instruments aboard *Luna 10* recorded each hit by a micrometeorite. Now, as the probe continued to circle the moon, the same instruments continued to record micrometeorite hits. As these comparative rates were read out on earth, there was general astonishment. *Luna 10* showed more than 100 times as many hits while circling the moon as while en route to it. In fact, during one five-hour circling period, the probe was hit by these tiny space bullets more than 50 times. Scientists were at a loss to explain whether the moon was passing through a strange concentration of interstellar micrometeorites or whether the rain of particles was strictly a lunar phenomenon.

As indicated elsewhere, the most astonishing thing about U.S. and U.S.S.R. space explorations is not how far apart the programs of each are but how close. The achievement of a lunar soft landing and a fully balanced orbit of the moon is a case in point. Both the U.S. and the U.S.S.R. had been planning and preparing for these events for at least five years, and yet the achievements, after many unpredictable failures and delays, finally came within the same six-month period. Both *Luna 9* and the U.S. Surveyor were more than two years behind their original schedules, yet one could almost believe they had progressed under the same set of managers and the same internal colony of glitches.

Surveyor, on its very first try, astounded everyone by soft-landing brilliantly on June 2, 1966, four months after *Luna 9* and in the same general target area, the Ocean of Storms. Both the similarity of general principles and the contrast in detail is pertinent, especially since, beginning with *Luna 9*, the Russians became as frank about their moon probes as they had become about their manned flights. It was now possible to learn from each other's successes and failures.

Originally, Surveyor was to have been launched toward the moon, like *Luna 9*, from a parking orbit. But this plan was abandoned when the hydrogen-burning Centaur upper stage produced an uneven record of attempts to start itself in space. It was decided to send Surveyor direct from its launch trajectory—a plan that worked as beautifully for us as the Russians' plan worked for them. The new hydrogen-burning Centaur performed brilliantly. Our guidance, like the Russians', was so accurate that our payload would have hit the moon, even without a midcourse correction. During this delicate maneuver, instead of aligning on the moon and sun, as *Luna 9* had done, Surveyor aligned on the very bright star Canopus. Then three separate rockets fired in sequence on command from earth. The resulting burn refined the target point. When Surveyor was about 60 miles out, it also aligned itself straight down and fired its retrorocket. Nearer the surface, smaller vernier engines took over and delicately lowered Surveyor to an impact scarcely greater than a drop from an office desk. Across the Ocean of Storms, *Luna 9*'s batteries, which had recently made history, were now dead forever. But *Surveyor 1*'s batteries were constantly recharged by its remarkably efficient solar panels. Before the frigid lunar night closed in nearly two weeks later, Surveyor had sent back 10,338 photographs, including some of the moon's surface in color. The way in which *Luna 9* and *Surveyor 1* compared most significantly was this: both were marvelously instrumented scouts from earth sent by curious men who were as yet too afraid and unprepared to send themselves across an ocean of space equal to the span of 80 Atlantic Oceans—a new ocean of space far more formidable than the Atlantic. Their being there at all was an astonishing coincidence. One might wish that these two brilliant substitutes for man's five senses, these adventuring robots fashioned by the two great rival cultures of earth, had somehow plunked down into the same terribly lonely crater where—in their common shell hole far from the insularities of their inventors—they could have shared their power and their intelligence and helped one another uncover the truth about the moon and, in the process, perhaps, about man himself.

This was the second extraterrestrial proximity of the brainchildren of rival societies. In late 1964, the *Mariner 4* Mars probe launched by the U.S. and the *Zond 2* Mars probe launched by the U.S.S.R. just two days later were approximately together on their competitive journey toward the red planet. *Zond 2* had a delayed start, but thanks to what the Soviets referred to obscurely as a "new control system using plasma engines," it caught up with *Mariner 4* at a distance of 3¼ million miles from earth. But what began as a spirited race dissolved as *Zond 2*'s communications again failed. Without radio tracking, there was no way to determine *Zond 2*'s position in relation to Mars, and it was its rival, the U.S. *Mariner 4,* that sent back the first close-up photographs of the moonlike surface of Mars.

Russia's long distress with its planetary probes was nearly compensated for in early 1966 when *Venus 2* passed within 15,000 miles of its target and the Soviets announced the landing of the automatic spaceship *Venus 3* on the cloudy planet. The impact, after a 106-day journey of more than 24 million miles, was a brilliant but slightly qualified success. The loss of radio contact just before the most critical period of the flight—descent and impact—caused a number of U.S. observers to believe that a soft landing had been attempted and had failed. Russian sources later revealed that "*Venus 3* was to have landed on the surface of Venus in the center of the visible disk of the planet. A soft landing on Venus was planned by means of a parachute system." Another credible supposition was that the expected fiery surface heat of 800°

F discovered on December 14, 1962, by the U.S. *Mariner 2* had simply been too much for *Venus 3*'s circuitry. The Russians, however, have never had much faith in *Mariner 2*'s temperature analysis of Venus.

The Russians announced they had made a successful midcourse correction of the 2,116-pound probe on December 26, about two months before impact. Once again England's Jodrell Bank director, Sir Bernard Lovell, made instant headlines by speculating that possible Soviet contamination of the planet by earth bacteria might have "endangered the future biological assessment of Venus." He had no information and was simply shooting in the dark. However, he did help prod the Russians—who had often criticized the U.S. lack of sterilization of deep space probes—into an admission that "the descending apparatus of *Venus 3* was sterilized before the start to destroy all terrestrial micro-organisms and to prevent their possible carriage to Venus."

Soviet guidance on *Venus 3* was superb, even though the last-minute loss of communication was a blow to science. The fact that one out of three Venus probes had actually made impact illustrated the Soviets' somewhat desperate shotgun approach, as well as their determination. Evidence indicates there were not just two, but three, other Venus attempts. An earth satellite the Soviets designated as *Cosmos 96* had been launched on November 23, 1965, just one week after the launch of *Venus 3*. This probe was believed by U.S. observers to be *Venus 4* when its inclination from the equator proved to be virtually identical to that of *Venus 2* and *3*. U.S. radar tracking showed that when *Cosmos 96*, or *Venus 4*, attempted to leave its earth-parking orbit an explosion of some sort occurred that broke the orbiting vehicle into eight pieces (compared with four tracked pieces for *Venus 3* and three pieces for *Venus 2*). Russian sources have never confirmed, however, that *Cosmos 96* was actually the third Venus probe launched within an eleven-day period. They obviously were determined to do the Venusian scene in the 1966 window.

Insofar as the moon was concerned, the year 1966 saw the U.S. and U.S.S.R. again involved in the near-incredible coincidence of basically similar lunar missions. Both countries repeatedly put probes into lunar orbits for the purpose of photographing the surface and, presumably, selecting primary and alternate manned landing sites. The U.S. Lunar Orbiter went into its circling mode on August 14 and carried out more than 6,000 commands, including the transmission of 215 highly detailed photographs, before it was intentionally crashed against the unseen side of the moon on October 29. Even as this moon satellite was called down, Russia sent up *Luna 12* to join *Luna 10* and *11* in orbit. Both *Luna 10* and *11*, because of tumbling, had been unable to transmit usable pictures, although they sent back valuable information on circumlunar space. Now *Luna 12* was sweeping in a wide elliptical orbit—ranging from 1,078 miles apogee to 62 miles perigee, the closest any Soviet orbiter had yet come to the moon. After four days in orbit, *Luna 12* sent back pictures of the lunar surface made with a special phototelevision device. The delay was caused by tumbling, which was finally corrected by a signal from the Russian control center.

Luna 12, a weird combination of cones, balloons, bell shapes, boxes, struts, and outboard thrusters, resembled an overburdened Rube Goldberg contraption. From the bottom up, it contained huge braking engines, four outboard attitude-controlling engines, gas containers for operating the astro-orientation jets, antennas, the optical-mechanical-electronic astro-orientation engines and system, chemical batteries, space radiator, instrument compartment, a phototelevision mechanism, and, finally, at its far pole, its radiometer. *Luna 12* had such a collection of engines and functions that considerable heat was generated during its operation. Because of this and solar radiation, what the Soviets call "a passive-active thermoregulation system" was installed to maintain a functional internal temperature. The system was aided by the application of certain colors, thermal insulation, and special screens on exterior surfaces. Nearly a true robot, *Luna 12* was one of the most advanced and most complex integrated vehicles ever launched by the Soviets into space.

By the end of 1967, Soviet probes on and around the moon were not only giving them a wealth of experience in lunar research, they also gave them almost the identical ability we had of selecting primary and alternate manned landing sites. Unlike the long-established moon experts in the U.S., Soviet selenographers and astrogeologists came to rather positive and unequivocal positions on the nature of the lunar surface quite early. As we have seen, they published their conclusion that the moon was basically basaltic and that it had a common origin with earth just after the flights of *Luna 9* and *Luna 10*. In contrast, U.S. lunar traditionalists were still debating these findings through the summer of 1967.

With the exception of Dr. William Pickering of JPL and the brilliant and enthusiastic Dr. Eugene Shoemaker of the U.S. Geological Survey, most American lunar experts were slow to react in terms of comprehensive theory to the remarkable new evidence obtained by close-up photographs and on-the-spot samplings of the lunar surface and surrounding space. At the same time, the main body of U.S. astronomers and cosmologists, who had been surprisingly aloof to, and occasionally disdainful of, the significant new tools of rocket astronomy, were still preoccupied with far-distant objects and the more esoteric phenomena of the expanding universe. Many of them, especially the traditional coteries at the California Institute of Technology, openly criticized the U.S. lunar programs. It remained for a few young men, chiefly Shoemaker, to design such new tools of astronomy as television cameras and stereopticon viewing devices and to use the new data thus acquired to formulate logical explanations of the nature and metamorphosis of the lunar surface. Shoemaker, for instance, early developed his highly significant studies of lunar stratification which led to a reasonable dating progression of visible geologic formations. Even so, it was not until the fall of 1967 that U.S. lunar experts finally agreed to a joint announcement that the moon was composed of basaltic rocks like those of earth.

As we have seen, for several years the Soviets had taken advantage of every favorable window for the exploration of the planets. Although they had a high percentage of failures, they were becoming quite candid about those failures, possibly from a sense of growing certainty that their objectives would eventually be achieved. In June of 1967, still another favorable window opened for a Venus trajectory and into it, both the U.S.S.R. and the U.S. inserted, almost simultaneously, two new competitive Venus probes. On July 12, without revealing the specific intent of their highly ambitious project, the Soviets launched their 2,433-pound *Venus 4* from their customary earth-parking orbit. It was the heaviest planetary probe on record. Two days later, the U.S. launched *Mariner 5* with the announced intention of collecting Venusian data during a 2,200-mile flyby pass. The 220-million-mile journey would take each probe about the same amount of time, just over four months. By mid-October, however, the U.S. probe had pulled within 36 hours of the Russian ship.

World knowledge of the planet Venus up to the late summer window of 1967 included no exact figures on atmospheric pressure. Estimates varied from 100 up to 500 times the 15-pounds-per-square-inch earth pressure at sea level. It was believed by some that the upper atmospheric strata of Venus becomes so hot on the sunny side that it ionizes and rotates to the night side, completely circling the morning star every four days. The temperature and nature of this atmosphere had long been the subject of speculation, and there were numerous other mysteries.

To prevent another irretrievable communications failure, the president of the Soviet Academy of Sciences, Academician Mistilav V. Keldysh, sent a cable to the director of England's Jodrell Bank Observatory, Sir Bernard Lovell, stating that the Soviet mission would be "of primary importance for the entire world of science." He expressed the hope that others might share the opportunity and responsibility of gathering whatever information *Venus 4* sent back.

At the time and inclination Keldysh specified (October 18, 1967: 0345 o'clock Greenwich meridian time; six degrees above the horizon), Sir Bernard expectantly adjusted his powerful radiotelescope to the specified frequency and aimed it at the dark sky. A strong two-toned signal immediately registered the Soviet probe in-bound for Venus. He held his fix and his rhythmic signal for one hour and twenty minutes. Then: sudden silence.

At this point, the one-ton probe was entering Venus' fiery and wildly turbulent atmosphere, and all signals were blocked by ionization. Sir Bernard, any more than the Russians themselves, did not know precisely what was transpiring. The Soviet plan called for the ejection of a round instrument package containing a tough, heat-resistant, braking-canopy device and surrounded by a thick hide of insulating and shock-absorbing material, the outer rind of which was about four inches thick. The sphere was weighted at the bottom and at its top was a dishlike radio transmitter that had to be aligned toward earth in order to project effective signals 50 million miles across the ecosphere. As this cradled and complex robot descended for its 90-minute ride toward what the Russians called "hellish fires," it was buffeted and swept by alien winds at velocities of hundreds of miles per hour. In the grip of such unearthly gales, it was impossible to hit a selected landing point. But somehow the coded signal of *Venus 4* came through.

Fifteen seconds after Jodrell Bank lost the original rhythmic, two-toned signal, a weaker and different set of signals began. As *Venus 4* settled through a 15-mile layer of scorching turbulence, the new faint signals continued. When they finally inexplicably ceased, Sir Bernard cabled Keldysh to inquire if further signals could be expected. There was no direct reply. *Venus 4* had perished, either just before or at the time it hit the surface, along with the boiling metal medallions and hammer-and-sickle emblems it contained.

The next day, Keldysh, who alone had the key to the code Jodrell Bank had intercepted, announced that his own interception showed Venusian cloud temperatures ranged from as low as 104° F. to as high as 536° F. The surface pressure it reported was 15 times that of surface pressure on earth. In addition, the Venusian interloper reported an atmosphere of 98.5 per cent carbon dioxide—the nonpoisonous gas on which green plants feed. Surprisingly, no nitrogen was reported. On the way in, it had found no trace of either a magnetic field or poisonous radiation bands such as those which encapsulate the earth. It had, however, discovered a very weak hydrogen corona above the dark side it penetrated.

Even before *Mariner 5*'s data began to come in, the Russian findings were more than enough to confirm that the planet nearest the sun from the earth was shrouded in inhospitable gases unlikely to support anything but high-altitude organisms at most. It was obviously much too hot to support any life as we know it, that could not be isolated by refrigeration from the incandescent volcano of its interior.

A day and a half after *Venus 4* made its plunge into the unknown, *Mariner 5* swept by at nearly 20,000 miles an hour some 2,400 miles from the surface. It sampled the atmosphere by radio waves, absorbing about two hours of data. Some of the data was transmitted back to earth live but most of it required over 30 hours to transmit from on-board tape recorders. Although there were discrepancies, Mariner's findings, in general, tended to confirm those of *Venus 4*. It estimated a carbon dioxide content of between 72 and 87 per cent and a surface pressure about seven times that of earth. On the day side of Venus, Mariner detected hydrogen particles some 10,000 miles from the planet and discovered a hydrogen corona on the day side just as bright as that found above earth. As Mariner flew behind Venus, its signals had to pass through successively denser layers of atmosphere. This not only provided a rough analysis of the atmosphere thickness (estimated somewhat less than 15 miles) but also revealed the possible presence of mysterious atmospheric substances or structures which may have been violent storms or dense layers of clouds. These undefined discrepancies were noted on both sides of the planet. Mariner also read out in the some 40,000 data units it transmitted some inconclusive evidence that a Venusian magnetic field was, indeed, present. U.S. spokesmen who thought they detected slight magnetic field variations were extremely cautious in contradicting the Soviet findings, however.

Although each probe tended to confirm and validate the other, there was no denying the superiority, the precision, and the boldness of the Russian undertaking. After numerous disheartening failures, they had made a brilliant and unexpected penetration of Venus. Their planetary program, long second to that of the U.S., had made an astonishing stride into the future. The U.S. had no comparable program. The two remaining Mariners the U.S. had scheduled would both be directed at Mars, and neither included plans for actually penetrating the atmosphere. The often-postponed U.S. Voyager program, which did envision a Martian soft landing, was in the process of being canceled even as *Venus 4* transmitted data.

Opportunities for favorable Venus flights occur once every nineteen months, and Russia had missed only one of these since 1962. The persistence and technical ingenuity that had finally paid off for them and which projected them to the forefront of planetary exploration had given them, above all, an enviable momentum.

15
Military Applications

ANY book or chapter that purports to discuss Russian missile strength should begin by saying that no one in the West knows very many specifics about the military rocket situation in the Soviet Union. This work is no exception. I have finally come to this conclusion, after dozens of background interviews in preparation for this book and for related articles in *Fortune* and other magazines. For several months, I talked to authorities in the Rand Corporation, Department of Defense, Library of Congress, Central Intelligence Agency, National Aeronautics and Space Administration, and the President's Space Council, in various offices from Washington to California. In the face of yards of published material that might, at first glance, indicate otherwise, I have reluctantly concluded that we are basically uninformed on the subject. I did not, in short, end up with a serene feeling of gratitude toward our peacetime, anti-enemy professionals.

Nevertheless, scarcely a month goes by that some "expert" does not pontificate on what the Russian generals are up to or—that old saw—where we stand in the space race, or who has the largest boosters. President Johnson brags privately that he knows how many missiles the "enemy" has. Each year publication in England of the latest speculations on the number and quality of Soviet rocket arms by the authoritative-appearing Institute of Strategic Studies—a sort of *Janes* of the rocket world—gets impressive front-page notice. The rocket magazines publish, from time to time, encyclopedias of military rockets together with pictures of the latest rockets the Russians elected to parade in Red Square. This catalogue is always accompanied by expert speculation about the characteristics, ranges, warheads, etc., of these alien rockets. Periodically, England's David Divine of the solemn London *Times* tells us just what the strategic rocket situation is in Russia.

Scarcely a week goes by that we do not recover a reconnaissance satellite launched from the Western Test Range in California (in 1966, there were 41 classified launches from the Western Test Range). So we know, at least, where some of the holes are in Russia. In 1966, Secretary of Defense McNamara, faced with the enormous cost of a nationwide, computerized anti-missile system, took the unusual measure of making public a statistical and revelatory portion of this strategic debate. Yet what it all shakes down to is the subjective impression that, under the mystic and shielding cloak of classification, lies a great mass of ignorance—an ignorance only recently slightly illuminated by remarkably clear pictures of the earth taken from orbital altitudes. Too often have I taken pains to absorb "vital" conclusions of experts of Soviet rockets—some of them with stars on their shoulders—only to discover, sooner or later, that the fundamental premise, as well as the stated buttress of their specifics, came from an analytical article in *Time, Aviation and Space Week, Newsweek, Technology Week, The New York Times, U.S. News & World Report,* or the *National Observer*. And in most cases it was relatively easy to see that the article, not the "expert's" opinions, came first. Most U.S. experts on Russia exaggerate their knowledge a great deal, possibly because, like Arctic explorers, they are quite jealous of one another. And many of them cannot resist the most vapid of all ploys of expertsmanship, the oblique suggestion that vouchsafed conclusions are based on privileged access to classified information. Of all U.S. Government agencies, I found NASA's representatives to be least informed about Soviet astronautics. Their frequent insight-connoting "No comment" to the press, I became convinced, concealed more ignorance and lazy or indifferent scholarship than real knowledge.

I recall what it seems to me are three of the biggest military rocket flaps that have occurred in the first decade of the U.S.–U.S.S.R. space age. The first flap was the reaction, discussed earlier, to the one-two-three blow of the Sputniks. The second occurred when a U.S. Air Force Chief of Staff was invited to Moscow to witness the May Day Parade and came back filled with self-importance and bristling with some solid ammunition for his strategic-rocket budget increase. The third occurred when Vice-President Nixon on a visit to Russia spotted some peculiar "dome shapes" on the outskirts of some Soviet cities. All of these occurrences resulted in profound soul-searching and policy adjustments on the very highest levels. Yet, quite obviously, the Russians were in cool control of the information revealed and obtained in each of these three situations. A powerful nation that has confidence in its reliable intelligence of the strategic strength of a rival powerful nation would not need to blanch each time the antagonist calculatedly tugged the curtain slightly more open.

I do not mean in any way to sound disrespectful of the vast governmental machinery for delving into Russian secrets. It is a massive, expensive, and in many ways impressive effort. But I just do not think in the field of military astronautics we are able to ferret out with even a massive effort as much as a third of what the Russians get from us simply by sitting back in an armchair and reading the national publications listed earlier. We get, I feel confident, very little the Russians do not wish us to get—and this includes information we get from our Midas, Samos, and Ferret reconnaissance satellites. The Russians are probably close to the point where, at their discretion, they are just as capable of closing off this source of information as they were of stopping the U-2 flights. But they have not done so, nor has there been any apparent Soviet regret over their earlier approval of an open-skies policy for space. Besides, the space age, coupled with infrared photography, has revolutionized the entire art of military camouflage.

MILITARY APPLICATIONS

Although the subject is highly classified, high-ranking U.S. officers and intelligence experts are so impressed with the high quality of our satellite reconnaissance—whose photographic versions can, from 300 miles up, delineate an object just 12 inches across—that they derive and communicate a strong sense of security. The understandable confidence of those who have been privy to these electronic military marvels, however, overlooks two significant factors: (1) the reason the Soviets originally permitted spy devices (by agreeing to an open-skies policy for space, when they could have opposed such a policy by a number of means) was so that they themselves could make use of all such intelligence-gathering devices, and (2) the Soviets can be just as adept at misleading a camera, an orbiting receiver, and an infrared sensor as we are. Says Dr. Ralph Lapp: "A hole in the ground outside a Soviet city is easily converted by our war culture into a highly efficient ballistic missile defense system. . . . It is curious how our cultists, otherwise so untrusting of the Soviets, believe them implicitly with regard to technological innovations."

Pride in advanced gadgetry sometimes exists only so long as we do not see the opposing gadgetry. And since this is something the Russians are not about to let us see, it is likely that certain privileged individuals will unconsciously reflect great pride for some time to come.

Having revealed this personal and highly intuitive lack of confidence in our ability to acquire, despite the Penkovsky papers, substantive rocket secrets from the Russians, there is still something to be said for taking a look at what they wish us to take a look at. There are some discernible patterns. The Soviet art of subterfuge, of course, is based more on the art of omission than that of commission, although both play a part. What the Soviets have committed to public and professional scrutiny seems largely to be three-year-old intelligence skillfully combined with three-years-away prototypes, and sometimes, we suspect, not even a Russian citizen himself can tell them apart.

First of all, we know that Russia also uses photographic reconnaissance satellites, partly because Khrushchev told us so, but mainly because we have observed their procedures to be so similar to our own. Again, similarity, not dissimilarity, is the most logical characteristic of the Soviet space program. It was widely assumed that the only purpose of unmanned Vostok and Voskhod flights was to man-rate these spaceships. Actually, after they were man-rated and the manned flights begun, the spaceships continued to be used as automatic vehicles in space which—*as far as we know*—were unmanned. As we have seen, when the Soviets adopted the broad designation of "Cosmos satellites" in early 1962, they announced the purpose to be broadly gauged, purely scientific, geophysical in character. One purpose of adopting the Cosmos label was, we now know, to conceal several military and paramilitary satellite programs. "Cosmos satellites" are launched from both Tyuratam and Kapustin Yar. The recoverable TY-Cosmos group began with 65° inclinations to the equator, but in 1964, to cover newer territory they were sent out at 51°. Their payload weight also increased so much that the original Vostok booster was finally superseded by a newer and more powerful one.

The Soviets seldom "read out" these satellites through telemetry. Data is collected and stored, and then the entire satellite and its secret cargo are commanded to re-enter and land within the borders of the Soviet Union. Many of these are known as "eight-day satellites" because they are observed to re-enter the atmosphere after precisely eight days before their orbiting booster re-enters. U.S. intelligence specialists, acquainted with the extremely high resolution of photographs taken by our own reconnaissance satellites, are convinced that Soviet photos quite easily have recorded a great deal more than merely every ICBM silo in the U.S. arsenal. They have flown over every U.S. county, not once but innumerable times, and under various lighting conditions. If the Soviets desired, they could, for instance, quite easily publish a census of the number and type of chimneys there are in the U.S. or the number of outdoor (and perhaps indoor) swimming pools, water wells, or telephone poles—or military objects.

We, of course, could do the same thing, a capability to

which Khrushchev gleefully pointed when he bragged about the quality of his pictures and offered to exchange satellite pictures with the U.S. President Johnson also acknowledged this capability to a group of Southern educators and government workers in a little-noted meeting in Nashville, Tennessee, in the spring of 1967. The President told his audience, supposedly off the record, that the U.S. satellite reconnaissance program alone had justified spending ten times the $35 billion to $40 billion the country had thus far spent on space. Because of these photographs, the President said, "I know how many missiles the enemy has."

Two significant factors in definitive and high-resolution satellite photography are a knowledge of forecast meteorology and the selection of the right time of day for photographing a specific target. A knowledge of meteorological trends is necessary, of course, because at any one time roughly a third of the planet is usually under a shifting cloud cover. In this forecasting, the Soviets make wide use of their meteorological satellites as well as ground-based computers. In the spring of 1966, Academician Laurentyev showed me a bank of M-220 computers in use at Akademgorodok, which, he said, could accurately predict weather twenty-four hours in advance anywhere in the Soviet Union. Both the U.S. and the U.S.S.R. can use orbiting photographic platforms in conjunction with computers to establish when a given area of the earth's surface is apt to be cloud-free.

The second consideration—how to get your camera in position at the proper time—is greatly aided by a new type of world chart that may become for space-navigation purposes what the familiar Mercator projection became for nautical navigation. The chart, whose Western version was first published in *Fortune* magazine in February, 1966, depicts earth-satellite paths over the earth's surface as straight lines. The linear display is possible only because lines of longitude are curved and land masses are drastically distorted. The advantage of this type of display is that you can select any photographic target on the earth's surface, assign it the preferred time of day you want your camera to appear over it (either midmorning or midafternoon to get best shadow definition), then plot a straight line to your desired launch pad. You can then quickly refer to the launch-site vectors for the precise orbital inclination and time of launch necessary to accomplish the objective. This would be true whether the objective be on the first, second, or any other orbit. By altering any of these variables, you can bring up your camera at any desired minute over any desired target, in some cases for more than a single pass within an eight-day period. But only rarely are Soviet satellite photographic missions for a single target. It is quite clear from a study of their shifting launch inclinations and changes in time of launch that they early developed a system of *continuous* coverage—as well as occasional spot coverage—of what they consider strategic world targets. Such targets have recently included geographic locations within Red China, as evidenced by the almost regular adoption of inclinations spread throughout the range between 51.8° and 72.9° (65° is considered the primary and preferred Soviet inclination for maximum and continuous surveillance of the U.S., including Alaska and Hawaii, and 91° is preferred fair-weather U.S. inclination for our classified polar-orbiting launches from Vandenberg Air Force Base, California).

Most U.S. observers estimate that at least 50 per cent of the first 140 Cosmos satellites launched were military reconnaissance satellites. Some U.S. experts believe that the Cosmos series, known as TY-65 (launched at Tyuratam at 65° inclination), are actually unannounced, manned Vostoks. Some observers also believe that the Cosmos satellites (*Cosmos 50* and *Cosmos 57*), which broke up or exploded in orbit into a number of pieces, were shot down from the secret Soviet anti-missile-missile site at Sara Shagan on the northwest shore of Lake Balkhash. It is known that the Soviets have devoted a great deal of emphasis to the possible destruction of what they consider aggressive satellites. Major General I. Baryshev, in an article called "Anti-Cosmic Defense," assumes that the U.S., like the Soviet Union, is quite capable of orbiting nuclear weapons. And, like a number of Soviet generals, he is worried over Red China's nuclear-space capability. Probably the first time the world will see

Luna 9 (centre) and a separate model of the probe which it soft-landed on the moon in February, 1966, on exhibit in the new Tsiolkovsky State Museum on Space History in Kaluga, which the author was the first American to visit. The photograph of the lunar terrain is one of a series taken by *Luna 9*. The tip of one of the petal-like feet of the probe is visible in the lower left-hand corner of the photograph.

The assembling of the third stage of a Voshkod spaceship, the first Soviet spacecraft to orbit multiple crews. A commodious and comfortable spacecraft, Voshkod, unlike its predecessor, Vostok, had no ejection system and was designed to land with its crew inside. Voshkod also utilized special landing or braking rockets, as well as parachutes, for the first time. The pointed nose fairing seen here is jettisoned after ascending through the atmosphere.

MILITARY APPLICATIONS   245

the Soviet anti-satellite system in operation will be in some cold war flare-up similar to the U-2 incident. If the Soviets felt their world relations would justify the deed, they are considered quite capable of shooting down or capturing (for display and propaganda purposes) one of our Midas or Samos reconnaissance satellites. Space voyages, no more than sea voyages, could not long continue without an intercept, rescue, or seize-and-capture capability. The only question is: which nation, in peacetime, wishes to be the first to demonstrate it?

The group of Cosmos satellites, known as KY-49 and KY-56, launched from the Kapustin Yar complex on the Volga River, have not been referred to in Soviet scientific literature and are presumed by some to be smaller satellites that are used for navigation by Soviet ballistic missile submarines and by other naval units. Both Academician M. V. Keldysh, president of the Soviet Academy of Sciences, and Premier Aleksei N. Kosygin had indicated by 1966 that Russia had an operational navigation system in orbit. U.S. observers are particularly puzzled and intrigued by the KY-56 Cosmos satellites, which are launched five at a time from a single booster.

The Soviet ability to put up simultaneously five payloads caused a considerable Pentagon flap because nothing worries Air Force officers more than the possibility of the Soviets putting up a group or chain of orbital bombs completely circling the earth. The reason for the concern is obvious. Bombs on ballistic trajectories can be tracked and the time and place of impact predicted, but the time and place of impact of orbited bombs and the new Soviet fractional orbital bombs which Secretary McNamara announced in 1967 cannot be pinpointed as easily. The U.S. is protected with the Dew Line and other segments of a network of so-called early-warning radar stations. Our entire intercept and counterstrike capability is based on early warning of at least twelve minutes, and, in most cases, fifteen minutes. Millions of dollars have been spent and thousands of bodies are employed to shrink each of those minutes. But, according to one informed intelligence officer: "If they put up a string of orbital bombs, any one could descend at selected times and targets and our vital early warning system would be useless. We'd have only four to six minutes, and we can't operate at present on that slim a margin."

The capability of both Russia and the U.S.—beginning in about 1965—to orbit nuclear weapons that could, on ground command, be called down to strike a desired target was to become the real sword of Damocles of the 1960s. Those scientists and researchers of the Manhattan Project who put together the world's first A-bomb in the mid-forties knew, of course, about the V-2 rockets that were raining down on London and Antwerp, but who among them could have foreseen that in twenty years the weapon they had created could be parked, or stored, above the atmosphere as the ultimate weapon of threat and intimidation? The threat would not have been less had Hitler tunneled under all of France to place and fuse strategic detonations. How the world got itself into that situation and how the two most powerful nations of the world felt compelled to debate and discuss the feasibility of a vastly expensive and otherwise useless antimissile defense system deserve a brief review.

When *Sputnik 1* went up in 1957, Russia already had a reliable, if bulky, prototype ICBM. The U.S. did not. There was then a definite missile gap. We were not able even to get our first Atlas off the ground until 1957, and it was November of 1958 before our first far-from-operational Atlas flew full-range from Cape Canaveral. In 1959–1960, at a time when we were still trying to make our ballistic missiles reliable, Russia organized its fourth military arm, the Strategic Rocket Forces, under Marshal Mitrofan Nedelin, who had been an artilleryman in the Spanish Civil War. The nearest U.S. equivalent was the Inglewood, California, Air Force Systems Command under Lieutenant General Howell M. Estes, a Bernard Schriever protégé. As the sixties began, these two rival commands and generals launched one of military history's most frenetic, desperate, and wasteful arms races. The confluence of advanced nuclear and rocket technology demanded—each nation felt—as many ICBMs and IRBMs as possible, as fast as possible—ready to fire, ready to retaliate, a million eyes for a million eyes. The phrases "massive retaliation" and "deterrent to global war" became catchwords in the command circles of each of two large and mighty nations who had never been at war with one another and who cared for thousands of veterans wounded by their common enemies.

In the spring of 1961, through a lucky accident, I was able to get a firsthand look at the rocket firing line of one side of the world's newest arms race. As a *Time* correspondent stationed in Chicago, I requested and was granted permission to visit and write about the most advanced underground bases for Atlas, Titan, and Minuteman ICBMs then under construction in the U.S. The overall project, I discovered when I landed at General Estes' headquarters in Inglewood, involved moving 37.5 million cubic yards of earth in order to convert 1,600,000 tons of steel and 2,700,000 tons of concrete into underground living, administrative, and rocket-launch facilities. The total cost of the project involving 20,000 construction workers in eighteen states was $7 billion.

I flew first to Great Falls, Montana, where 150 Minuteman silos were being sunk in an area twice the size of the state of Maryland. On rolling, dun-colored hills, huge earth wounds 90 feet deep would suddenly appear. One of the rocket locations was within sight of a group of about forty elk grazing on a small forested peak. And everywhere, across undulating wheat fields, over arroyos, through meadows and pastures, snaked long gashes in the land that were to carry the interconnecting power and communications cables.

The next day I flew to Colorado, where five men had just been killed when a 58-ton Titan silo door fell on them. Deep within the ground, I walked down a long, cool, yellow steel tunnel to the commodious and cheerfully painted control center. Here each circular room was insulated against vibration by a heavy rubber collar one foot in diameter. To soften the earth jar and massive vibrations of a nuclear-rocket explosion, even the toilets were mounted on rubber; the plumbing conduits leading to and from them were of flexible rubber and nylon. I also went down to the bottom of the silos themselves—a strange world of damp concrete, serpentine cables, and intricate and freshly painted rocket fueling and launching tanks, umbilicals, and skids. These dank holes, more than 100 feet down, were too deep even for mud daubers to build their earth-colored nests. The next day I flew to Abilene, Kansas, to see the big, bright Atlas holes that had been punched down in feed lots and cow pastures. There were no rockets yet in these Atlas silos, and I remember the strange doors that led to the cavernous and empty shafts. Because these doors opened into nothingness and sure death, they were painted bright yellow and their door knobs—as an additional reminder of what was on the other side—were located in the lower right-hand corner about six inches from the floor.

This lonely five-day barnstorming tour through the space age's underground and the reflections it prompted each night in some strange motel room were a sobering experience. Looking back on it now, the death tubes in the earth, fortunately and objectively, can be referred to as instruments of "the war that never was."

It is not known what silos or other ICBM launch installations the Russians had in 1961, but if "the war that never was" had actually happened, it would undoubtedly have been the most bizarre, complex, and futile human conflict that was ever devised by man in his centuries-old and evolving effort to put something besides his bare flesh between him and his assumed enemy. In the first place, few of the officers briefing me had much faith in the ability of our rockets to get out of the ground safely with a nuclear warhead, much less find their transoceanic or transarctic way to target. "Those spring-mounted platforms and rubber cushions," one lieutenant colonel said privately, "are to protect one silo from its neighbors, not from the Russians. If one out of three just gets out of the hole, we're lucky." In all probability, the Russians were having the same sort of troubles, and if the two nations had actually shot at each other, they probably would have done more damage to their own launch facilities and home personnel than to distant geographical targets.

In the second place, "the war that never was" would not have known how to end itself unless it ended automatically out of mutual frustration. A look at the prevailing military doctrine of each country shows an alarmingly naive position. The most frequently repeated—and therefore believed—war prognosis in both Washington and Moscow was that nuclear-rocket warfare—now that it was really here—would be over in a matter of hours, even, perhaps, minutes. Khrushchev often used this argument against his Red Army generals. And we were so convinced of it we did not even bother to build in a repeat capability in those expensive concrete cannons we were sinking. All of our cannon were one-shot cannon. We did not even build an underground missile storage area for possible and eventual reloading. The reason for this, anyone would tell you, was that each country would annihilate the other practically on the very first exchange. This is perhaps one of the most ridiculous doctrinal premises ever perpetrated on two supposedly civilized countries.

For speculative purposes, let us see how the war of 1961–1962 may have been fought. For some reason or other, let us say the Strategic Air Command of the U.S. and the Strategic Rocket Forces of the U.S.S.R. both got the signal to push the button on November 11, 1961. Both countries attempt to fire their first salvo of fifteen ICBMs. The hard fact of the U.S. law of averages for rocket launches actually gave us 29 successes out of 41 launches in the year 1961, and the Russian average was probably only slightly more favorable than ours, due to their longer experience and more reliable boosters. So out of our first fifteen, at least five abort in the silo, doing extensive damage to the areas around, for instance, Great Falls, Denver, and Abilene. The same thing happens in Russia. It is a fairly safe assumption that another five abort in the air, landing in the oceans or on the ice pack. Five, say, of each country hit the enemy, wiping out, possibly, two large cities and two small cities, including three ICBM launch pads.

It would be really difficult to imagine the rocket military-command confusion in this situation. Remember, this was in the days before reconnaissance satellites. Who shot first? Did we do more damage to ourselves than to the emeny? Did the enemy have errors too? Can we force him to fire again so he will hurt himself even more? And most important, what do we do next? Probably nothing. Or perhaps somebody may give the order to shoot only at enemy ICBM launch facilities. This is a primitive tactic, militarily speaking, the old counter-battery fire of World War I—only our sites are hard—the doors alone are five feet thick and weigh 58 tons. Russian rocket troops do not have as many hard sites as ours do, but their ICBMs are highly mobile and we do not know where they have moved to. It would be extremely difficult to knock out enemy launch capability completely no matter how imperative it may seem to do so. Each side gets off fifteen more ICBMs with approximately the results of the first salvo. But is the war over? Not by a long shot. Over 95 per cent of each country is untouched except by radiation, and the main radiation danger is temporary. Millions of people go underground to wait out the war. And in 1961 there would be only about enough armed ICBMs left for one or two more self-damaging salvos. Who would give the order to fire now? And at what? Remember the rubber-mounted toilets would still flush deep within the Colorado foothills. Food rations could last as long as four months. Nobody has to come up.

I think it is clear that it was probably best for all concerned that the War of 1961–1962 never was. Some generals would have been demoted for promoting some bad doctrine, and there would have been no heroes, except perhaps the Space Age Plumbing Company of Muncie, Indiana.

In fact, the more one thinks about it, the more inconvenient it would have been to have such a doctrinal war any time in the 1960s, even when the ratio of damage done at destination to damage done at point of origin worked in favor of both sides, and as the anti-missile efficiency reduced anti-efficiency factors.

The sixties produced interesting milestones in the military area. In 1960, with its new Strategic Rocket Forces in being, Marshal Andrei A. Grechko was neither rebuked nor contradicted when he announced to the Supreme Soviet that the rocket arm was "actually the main branch of our armed forces." In December of 1961, the U.S. and U.S.S.R. supported a unanimously adopted resolution for what came to be known as "peaceful uses of space." The resolution established the principle that space and celestial bodies are free for exploration and use by all nations and not subject to expropriation by any nation. Russian support of this "open-skies policy" for spacecraft, if not U-2s, greatly surprised us because we had assumed that Russia knew more about what went on in the U.S. than the U.S. knew about what went on in Russia and that Russia, therefore, would not want our cameras and satellites orbiting over U.S.S.R. territory. Besides, U.S. maps available to the Russians were more accurate than Russian maps available to the U.S., and we were tremendously handicapped by not knowing the precise location and distance to any given target. It is a major puzzle to this day why Russia decided on mutual permissiveness regarding spy-in-the-sky satellites at about the same time they gave Marshal Nedelin a medal because one of his rockets shot down Francis Gary Powers' long-winged U-2 over Russia. Possibly some Russian general used the argument that manned U.S. satellites did not pass over Russian territory, while manned Russian satellites passed often over U.S. land masses. But they would have known this was a temporary situation, since it is unlikely that the U.S. Air Force would confine its $1.5 billion Manned Orbital Laboratory to non-polar orbits. (Due to the comparatively low latitude of the U.S. bases and launch safety requirements, only U.S. polar orbits can fly over Soviet territory.)

The U.N. Resolution, however, was mainly a statement of good, and somewhat wishful, intentions; it did not, for instance, permit policing or inspecting orbital objects. Perhaps with this thought in mind, about a year later, in September of 1962, U.S. Deputy Secretary of Defense Roswell Gilpatric announced with a straight face: "We have no program to place any weapons of mass destruction in orbit. . . . An arms race in space will not contribute to our security. I can think of no greater stimulus for a Soviet thermonuclear-arms effort in space than a U.S. commitment to such a program." Gilpatric added, however, "We will, of course, take such steps as are necessary to defend ourselves and our allies, if the Soviet Union forces us to do so."

Russian spokesmen at the time were not listening too well, for they were engaged in the same sort of dual dialogue and did not trust it any more than we did. "Our program is peaceful," said Khrushchev time and again, and subordinates repeated it, from marshals to cosmonauts.

David Divine of the London *Times* estimates that by 1964, Russia had about 200 ICBMs poised and ready, plus about 650 highly mobile IRBMs. The U.S. had only slightly more ICBMs operational in the Strategic Air Command, but mass production would soon dramatically increase our lead in number of ICBMs only. In 1963 Secretary McNamara had already announced, with permissible exaggeration, that we could shoot down any object in orbit. But 1964 would have been a bad year for a war also. For one thing, we had never in our lives fired a strategic nuclear rocket in anger from that feed lot outside of Abilene, and our untested and unproved anti-missile-missile—if enough could have been nuclear-tipped on time—would probably have done more damage to the U.S. or our neighbors than our ICBM dogs would have done. Russia, also, was not very well checked out on these new defensive gadgets. This same year President Johnson let it be known that our radar could detect objects beyond the horizon, but a rocket war would not have gone well in 1964. By then most American bomb-shelter companies had gone bankrupt, and my Government friend, Virgil Couch, to whom I had referred in *Time*'s cover story as "Mr. Civil Defense," was having a hard time getting industry to build its cellars big enough or deep enough. In Russia, Muscovites praised the foresight of the designers of the world's most beautiful subway—located 300 feet beneath the ground—and the designers of the subways in Leningrad and Kiev, also very deep. Russia's Civil Defense Director Marshal Vasily I. Chuikov dug few new holes.

By 1965, Russia had announced that it already could place nuclear bombs in orbit. Said Soviet Colonel-General V. P. Tolbubko: "Powerful missiles are being created that can ensure delivery to the target of nuclear warheads both on ballistic and orbital trajectories and that are capable of maneuvering within that trajectory." The prevailing doctrine in the U.S. was that Russia, partly because of the expense involved, would not put nuclear weapons in orbit even though we could not then verify the ordnance nature of orbital objects.

The Soviets backed up their claims, in part, by twice parading the 115-foot, three-stage, segmented, "orbital" Scragg rockets in Red Square. Following the latter occasion—the parade on the 48th anniversary of the Russian Revolution—Soviet spokesmen reiterated the power and accuracy of their "orbital" rocket—statements already in conflict with the wording, as well as the spirit, of the October 16, 1963 U.N. Resolution. The announcement of the Soviet orbital bomb, from a propaganda viewpoint, was a blunder of the first magnitude. From an economic and strategic viewpoint, however, the Soviets obviously wished the U.S. to commit itself to the tremendous expense of an expanded Nike-X anti-missile system just as soon as possible. Even though the Nike system would have a better chance than our Sprint against one of their high-firing "overkill" bombs, the Soviet militarists obviously wanted another national U.S. financial commitment to go along with the urgent priority and economic drag of Viet Nam. This, of course, would hamper offensive rocket weapons.

By the end of 1965, press reports credited the U.S. with about 1,630 ICBMs, plus another estimated 656 nuclear-submarine-launched Polaris missiles and nuclear bombs deliverable by Air Force planes. The U.S. believed its numerical lead had now reached 4 to 1, thanks to crash, mass-production programs. England's Institute of Strategic Studies, however, said the U.S. lead in numbers of strategic missiles had decreased by 25 per cent during 1965. ISS also warned of a new Russian ICBM with an estimated warhead power of 30 megatons plus. A megaton, it is sometimes helpful to remember, is equal to the explosive force of one million tons of TNT. The 30 megatons of Soviet overkill were designed, we thought, to be fired at extremely high altitude to foil our defense. The U.S. numerical lead, said ISS, instead of

4 to 1, was more like 3 to 1 since we had now already scrapped our Abilene Atlases as obsolete and were soon to phase-out all of our expensive Titan holes in Colorado. At the same time the Soviets had upped their ICBM production by 40 per cent. For both sides, 1965 would have been an inconvenient year in which to have had a rocket war.

The next steps continued the world paradox, the insecure alternation of the ancient olive branch and the new arrow of space. While the U.S. spent $20 billion a year in Viet Nam, Russia spent an unknown amount on strengthening its rocket arm and aiming for a sudden spurt in the arms race. By 1966, Russia had demonstrated the multiple-warhead technique, partially revised its civil-defense program, begun deployment of its computerized anti-missile-missile, and publicly indicated its approval of a historic new space treaty. With the increasing belligerence and rocket-nuclear capability of Red China, Russia, for the first time, had to consider the possibility of a major enemy other than the U.S. and, in space-age terms, a potential IRBM enemy instead of an ICBM enemy.

The state of military rocketry reached the point almost simultaneously in both the U.S. and the U.S.S.R. where the most economic expansion of firepower could be derived from the development of multiple warheads, or the rocket bandolier. Five or more dispersed warheads would not only assure better and more target coverage, but would also confuse and complicate defense techniques and could be grouped in unpredictable arrays to overwhelm defense in certain desired areas. The great chess game was being played with even pawns of pure uranium, or so it would seem to budget directors in the second decade.

Now that Red China had an A-bomb and was developing rocket-delivery systems, Soviet Civil Defense Director Marshal Chuikov found Russian factories and other enterprises somewhat more interested in erecting shelters and training civil-defense personnel. He wrote a widely published article titled "There Is a Defense Against the Nuclear Weapon." Marshal Chuikov found it too early to substitute another word for the hate-mongering label "imperialist."

MILITARY APPLICATIONS

"If nuclear weapons are unleashed by the imperialists," he said, to be followed by a familiar Washington predicate, "there will be no distinction between the front line and the civilian rear." Marshal Chuikov urged each citizen to take specific steps in his own defense, but apparently the Russian individual was no more prepared to dig than his U.S. counterpart. Raymond Anderson reported a current Moscow anecdote to *The New York Times*:

> Q. What should I do if there is a nuclear attack?
> A. Pull a sheet over your head and start walking very slowly toward the nearest cemetery.
> Q. Why very slowly?
> A. To avoid causing a panic in the streets.

The most debated military question at the end of the first decade of the space age was whether or not—in either Russia or the U.S.—to deploy an expensive, computerized, coast-to-coast missile-defense system that would strike at ballistic, orbital, and fractional orbital bombs. The facts of rocket warfare were such that international anxieties had caused offense to progress even faster than defense, and both offense and defense had undergone far more progress than human intelligence could cope with. The facts also suggested the oldest economic saw in military history, but with a terrible new twist—that all those huge expenditures would be wasted unless we had a war; the new twist was—unless we had a war of annihilation, because rockets were now efficient enough for fallible man to do serious damage to his prospects for survival as a species. If we did not spend money, in other words, we might be annihilated. No man sentenced to the gallows has ever gone on a diet.

Secretary McNamara first mentioned the Soviet deployment of its ABMs on November 10, 1966, at the President's ranch in Texas. Several times earlier, U.S. experts had thought Russia was digging in its rocket-defense system but several times also they thought the Soviets had apparently stopped, probably for lack of funds. "There was now," Mc-

Namara said, "considerable evidence that they are deploying an anti-ballistic missile system."

The Secretary did not elaborate, but intelligence experts in high places believed the training center for ABM crews, as discovered by reconnaissance satellites, was at Sara Shagan on Lake Balkhash. The new desert base was so positioned relative to the Tyuratam complex that it could practice on both orbital traffic and incoming ballistic traffic. Sara Shagan was about 490 miles from the Tyuratam pads and about 300 miles from the new pads at Karsakpay. (It was not considered, at first, to be operating in conjunction with the new Soviet rocket base at Plesetsk, north of Moscow.)

Whether he intended it or not, McNamara's public statement on Soviet ABM deployment resulted in renewed pressure from Congress for a massive Nike-X program. This may have been just what the Russians hoped to bring about by their otherwise unseasonable boast about their orbital bombs and by their excavations around Moscow, Chernovtsy, Leningrad, Arkhangelsk, and Karaganda. Senator Richard Russell, influential chairman of the Senate Armed Services Committee and Defense Appropriations Subcommittee, asked for Nike-X production. Senator William Proxmire of Wisconsin and Senator Strom Thurmond of South Carolina, a member of the Senate Armed Services Committee, both appeared to lean toward the most expensive defensive military program in the nation's history. Senator Proxmire did not want to risk "temptation for a nuclear attack," and Senator Thurmond, who said he had had evidence of Soviet ABMs for two years, said further U.S. procrastination on Nike-X could be disastrous. The Institute of Strategic Studies whose annual exposures had really demonstrated and perpetuated a stalemate situation said, as if in relief, "no ABM system is at present capable of upsetting the strategic balance."

In the midst of this portentous debate on the arrows, another great debate took place on the supposed side of the olive branch. More than sixty nations were involved in the 1966–1967 treaty to limit military activities in outer space, but it was really a treaty of somewhat vague intentions between the only two spacefaring nations—the U.S. and the U.S.S.R.

MILITARY APPLICATIONS

It proposed that weapons of mass destruction (i.e., nuclear) be prohibited from orbit, on the moon, or on other celestial bodies. It also prohibited all military installations and maneuvers on the moon and planets. In effect, the treaty appeared to have the high ethical basis of one of the most successful international agreements ever signed, the amazingly applicable Antarctic Treaty of 1959. But the fine print was not much more specific than the Resolution of 1963. It did not, for instance, prohibit the orbit of spacecraft without "large" weapons or the use of space for unmanned reconnaissance satellites. Nor did it provide for inspection or flyby analysis of orbital spaceships or space stations to be sure no nuclear weapons were in cargo. Like most treaties that get signatures on them, it did not prohibit anybody from doing things they were already doing nor did it prevent anyone from making "contingency" decisions and preparations in case the other side ruled later to chuck the moralistic restrictions.

But, as 1967—the tenth space year—began, it was one of the good moments in space thinking, although it made bigger, if inconspicuous, hypocrites out of a great many people on both sides of the titanium curtain. Our Air Force-staffed and military-motivated Manned Orbital Laboratory project went ahead full steam, as presumably did the Soviet program for making a joint military craft out of the advantages and flexibilities of Voskhod and the 13-ton Proton. NASA, fighting for its budget life, smoothly changed its line from "purely scientific" objectives to what it was contributing and could contribute to our military objectives in space. The U.S. also went quietly ahead improving its computer-based analysis of the "mass" of orbiting foreign space vehicles, an analysis, many felt certain, that could distinguish a nuclear device in orbit. You must have one to know one.

By the tenth year, both powers also had made a seldom-mentioned accomplishment that served to extend the degree of the stalemate. Both had military communications satellites. Presumably the Russians' military communications satellites (Redmilcomsats) worked as well as those (Bluemilcomsats—or Transits) we had been launching from Vandenberg for several years. Presumably also, however, the antimilcomsat programs of both countries were still unreliable).

Not mentioned to the Russian public, but of even more space-age importance militarily speaking than orbiting hardware was Soviet progress in developing a chemical serum against radiation and radiation-resistant flora and fauna for use in long-duration space flights. An effective anti-radiation serum would have an incalculable effect on the future of manned space flights—especially flights to the planets. But it would have even more effect—in terms of billions of dollars—on civil defense and on the whole complex problem of nuclear warfare defensive strategy.

The bulk of this crucial investigation work in the Soviet Union has been carried out in Moscow's Radiation Genetics Laboratory of the Institute of Biophysics of the U.S.S.R. Academy of Sciences. The work is headed by space-genetics expert Nikolai Petrovich Dubinin, who has also traveled to the Urals, where he has engaged in studies of birds that so far have resulted in two books.

An obscure interview with Dubinin by Vladimir Gubarev in December of 1964 was inexplicably allowed to appear in—of all places—a Soviet weekly economics newspaper, *Ekonomicheskaya Gazeta*. In it, Dubinin said, in summarizing Soviet investigation of life in space: "It is precisely cosmic radiation that most dangerously affects both man and plant organisms. . . . Our first-ranking goal today is to produce on board space ships conditions which would enable human beings to carry out prolonged flight."

Then Dubinin adds significantly, "man in space can be protected against the action of radiation with special screens, helmets and other facilities. However, chemical protection can also be used. . . . Several chemical preparations have already been synthesized which, when introduced into the organism, protect it from the effect of radiation: they serve as a unique 'chemical shield' which assists the physical means of defense in coping with space radiation."

In addition to developmental work on this "chemical shield," which several Soviet scientists have admitted was a subject of high priority in the Soviet Union, Dubinin and his laboratory workers are involved in developing a strain of the plant Chlorella, which is resistant to radiation. Such a plant, with what Dubinin calls "radiation stability," would be invaluable in producing man food in a closed ecological system or space garden during the coming long-duration flights.

In developing their serum against radiation, the Russians are known to be using numerous orbiting test specimens from mice to Drosophila. Some of the specimens get their radiation doses and anti-radiation by accompanying cosmonauts in orbit; others, for comparative purposes, are irradiated on earth.

The most significant test of the new radiation serum is believed to have occurred on the February, 1966, orbital flight of the two dogs Veterok and Ugolyok into high-radiation zones. A specially designed probe was known to have been inserted in one of Veterok's arteries for the introduction of what the Russians called only "pharmacological agents." These agents were believed to include the highly important serum that, among other effects, caused Veterok's heart rate to increase. If the Russians are successful with such a serum, it will be of far greater consequence than the perfection of a new booster or a new superspacecraft.

Also seldom mentioned in public documents but of great importance to both countries, as well as another mutual augmentation of the stalemate, was the research and development work done on the real second-generation spacecraft—the maneuverable, re-entry orbital vehicles known as lifting bodies or wingless wedges. A lifting body—shaped somewhat like a bullet sawed in half lengthwise—as its name implies, relies on its broad body shape, instead of wings, for lift. It can maneuver in space with reaction jets, re-enter the atmosphere, and maneuver sufficiently at low altitude to land safely on runways. Both the U.S. and Russia are working on them. In the U.S., since pilot Milton Thompson flew the plywood, prototype lifting body, the M2FL, to a safe landing from 12,000 feet in 1963, both NASA and the Air Force have been working with models and full-scale craft. Thor missiles at Cape Kennedy have ferried up four-foot models that proved they could re-enter the atmosphere without being destroyed. The next eight-foot model, the SV5D, was launched atop an Atlas at Vandenberg, as both NASA and the Air Force began to get very excited indeed about their ability to launch a truly advanced maneuverable re-entry vehicle by 1970, atop the powerful triple-barreled Titan IIIC. In fact, the MRV looked so promising that some NASA engineers wished it could have been substituted for the lunar vehicle in Project Apollo. In Russia, all research papers, speeches, announcements, photographs, and drawings on lifting bodies simply disappeared from sight, beginning about 1962. This is usually a sure sign that the subject is becoming too hot for light.

The reason is obvious. Lifting bodies represent the perfect merger of the military aircraft and the military spacecraft—whether manned or unmanned, although their ultimate use is expected to be in manned versions. The ability to maneuver in the atmosphere means controlled and variable entry into space and affords a relatively wide choice of landing points. The ability to maneuver in space means many things: intercept and inspection capability, astronaut-rescue capability, seize-and-destroy capability, space-salvage capability. Such bodies have an important scientific rationale, but it is the militarists of both West and East who pushed for them. General Bernard Schriever, for instance, came out of retirement briefly in 1966 to say: "What I'm most unhappy with is the slowness in getting on with maneuverable re-entry spacecraft progress."

The shape of the payload of the future is that of lifting bodies. It will be the sleek, superbly insulated, multiple-motored, wingless wedges that will make the Vostok cannonball, the Mercury and Gemini, Voskhod, and Soyus canisters, and the top-shaped Apollo all look like well buckets, and will make gangling, boxy LEM look like an old-fashioned wooden wall phone. A wingless wedge that can pull in its space protuberances like a tortoise and prepare for its slippery, fishlike plunge into the atmosphere will be the look of the future in space.

If lifting bodies are the payload of the Soviet and the

U.S. future in space, the nuclear rocket is the booster to which lifting bodies will be mated. Little has been heard in the Soviet Union of nuclear rockets since the executed Soviet Colonel Oleg Penkovsky reported that the death of Marshal Mitrofan Nedelin (attributed to an aircraft accident) was actually caused by a nuclear rocket that exploded on the pad in the early sixties. And since the few papers on the subject, by Igor Morokhov and others, disappeared in the late fifties, the subject of nuclear propulsion, therefore, is considered to be one of great emphasis in Russian military centers. It has often been proposed in early Soviet papers as the vehicle to take Russians to Mars and Venus.

Very little has been said in this chapter about the size of Soviet military boosters. Booster size, like the "kill" claims of both sides in modern air warfare, is the most hackneyed aspect of the Soviet space program. Editors apparently will run shallow speculative stories about Russia's past or future boosters again and again. It was only the first two Soviet boosters—each with just over a million pounds thrust—that constituted a very real missile gap for about three years. Now that the boosters of both the U.S. and the U.S.S.R. have well exceeded 5 million pounds of thrust, all discussions of booster capabilities, for military comparative purposes, are useless and academic. There is no military purpose in space for which existing boosters are not adequate. In fact, boosters have become somewhat like elevators in Hilton hotels. If Mr. Hilton would like to move fifty people a minute in express elevators to the top-floor banquet hall, he has a capability to use one very large elevator or to use two or more smaller ones. So it is in military space. If the Russians want to orbit less than a platoon (or its equivalent in instruments), they can use the improved Proton booster; if they want to orbit a platoon or more they can use several rockets and assemble a commodious observation station in sections. At a time when the U.S. Air Force is having trouble figuring out a legitimate military purpose that needs to utilize the full 7½ million pounds plus thrust of our *Saturn 5* rocket, it is likely to be a long time, indeed, before any Russian general has a call for a larger booster. The first need for the 10 million pounds plus booster will probably be for sending man to Mars or Venus, and this will certainly not be a military mission between earth combatants. The first Russian to ask for 10 million pounds of thrust or over, is apt to be a scientist aiming for the planets.

The frantic, incredibly expensive, and self-canceling arms race in space, as indicated earlier, has finally brought to the profession of arms the enervating luxury of button pushing. We have the means of unleashing robot instruments of destruction on an enemy far too remote to see the whites of his eyes. Mine enemy grows older. Like myself, he likes his comfort. To broaden the already tremendously wide gap between his flesh and mine, both of us have constructed luxurious hemispheric contraptions. Rockets have become technological mercenaries. If one of the great, breast-plated sword warriors of history, such as Leonidas, were to return to survey the state of his ancient profession of arms, he might well spit and proclaim, "Cowards, all!"

Global war has been outmoded since the Day of Trinity in July of 1945, but this knowledge and this fact is slow to seep into human consciousness. Technology outraces man's comprehension. And man's traditional apprehension is still an unthinking man's reflex action. The insulated and unarmored button pusher on both sides of the titanium curtain is still the victim of his anxieties, his fears, his insecurities, and his inherited habits. In his fearful ignorance, he seems quite willing to mine the void and fuse the moon.

"By 1967," wrote the noted historian, Arnold Toynbee, "the whole surface and air-envelope of the globe had been knit together for purposes of warfare." It was quite clear, as the first decade ended, that this unity was achieved by the militarists of each of the two great warfare states. For better, or more likely, for worse, it was this insulated but strangely similar war corps that dominated public policy. But the ranks were elusive, for members were not simply a group of insentient warlords who gloried in power and might. Some were captive to the momentum of a chariot that was already racing when they were born. They would follow any order— be it to dismantle the stars. Some thought while acting and doubted, but acted. Others made unprecedented gestures of restraint. A case in point was the somewhat awkward but wise U.S. plea to the Russians to desist from suspected plans of deploying their anti-missile-missile system. It was as if we were saying: "If you won't, we won't, because neither of us can afford either the cost of deployment or the cost of ultimate employment." The enlightened search was for needed rapport, not appeasement.

The overwhelming impression was that each great antagonist was so preoccupied with his own magnificent and evolving gadgetry that he had little time to reflect, to attempt to comprehend the futility and ridiculousness of the ultimate and unending arms race in space. And so esoteric has become the science of warfare that few could debate or make clear what was happening.

Fortunately, George Orwell was not the last prophet. With prudent foresight, President Eisenhower in his Farewell Address warned both the U.S. and the U.S.S.R.: "In the councils of government we must guard against the acquisition of unwarranted influence, whether sought or unsought by the military-industrial complex. The potential for the disastrous rise of misplaced power exists and will persist." The outgoing President wisely cautioned: "In holding scientific research and discovery in respect, as we should, we must be alert to the equal and opposite danger that public policy could itself become captive of a scientific-technological élite."

In an age in which Dr. Ralph Lapp estimates the world has manufactured and stockpiled for possible war use the equivalent of 10 tons of TNT per human being, the biggest danger to survival would seem to be those nuclear-rocket militarists of *both* countries who are quite capable of promoting and endorsing unlimited "preparedness." In an age in which we possess an incredible overkill, it would be a bitter paradox, indeed, if both the U.S. and the U.S.S.R., who have thus far each been satisfied with a stalemate, felt it necessary to fully and expensively reconstruct that stalemate in the unlimited ocean above the skies. The horror is that each could prefer to do that to finding out what it is that its enemy is really afraid of.

"Isolation," said former Defense Secretary Robert A. Lovett

at West Point in 1964, "is what creates the real problem." Three years later in a notable speech titled "The Weapons Culture" at Purdue University, Dr. Ralph Lapp reiterated: "The power of isolation enhanced by insulation can be formidable . . . Karl von Clausewitz' dictum: 'War is an act of force, and to the application of that force there is no limit,' needs revision in the age of the megaton."

It is far easier and much less expensive to revise man's view of his fellowman than it is to revise the nature of war. When man's cultural isolation is removed, the fears of his being require less military insulation.

16
Death in Space

FOLLOWING the first two highly successful Voskhod flights, it was widely supposed that Soviet spacemen were in an excellent position to launch an ambitious flight series with their commodious and versatile new spacecraft. *Voskhod 1*, weighing 11,700 pounds and carrying three cosmonauts, was launched October 12, 1964; a little over five months later, on March 18, 1965, the launch of the 12,500 pound *Voskhod 2*, carrying space walker Alexei Leonov, was greeted as the second of the series. Since Voskhod was capable of holding three men and was about the same weight as the 12,000 pound U.S. Apollo, some observers believed the Soviets to have, in effect, bypassed the intermediate, two-man Gemini-sized spacecraft. With auxiliary components and propulsion units it was believed that Voskhod was even capable of going to the moon; especially since *Voskhod 2* contained a built-in tunnel for oxygen saturation and egress.

But as the new U.S. Gemini series met and extended its fast-paced flight schedule, it appeared that there were inexplicable delays in the expected Voskhod series. Gemini went on to complete brilliant rendezvous and docking missions and to demonstrate versatile maneuverability; and still the cosmonauts did not fly. Unaccountably, in fact, the entire ten flights of the Gemini series concluded in late 1966 without a single announced Russian manned flight—in Voskhod or in anything else. In Gemini the U.S. had completed ten rendezvous experiments and had docked nine times. By any logical calculations, this vital experience thrust the U.S. ahead in manned flight precursors to flights to the moon.

By early 1967, the continued failure of the Russians to accumulate manned-flight experience in advanced spacecraft had become a major and baffling mystery. Did this mean that the Russians were bypassing cosmonaut rendezvous and docking and planned on going directly to the moon on a stupendous booster even larger than the U.S. *Saturn 5*?

If so, what of their long nourished hopes of assembling a large space station in orbit; a feat requiring delicate orbital navigation. The enigma was basically unanswerable in terms of Western logic. Even the death of Sergei Pavlovich Korolev in January of 1966, a demise which at last lifted him personally and officially from the ranks of the anonymous space pioneers to the holder of the revered title, "Chief Spacecraft Designer," shed little insight on the mystery. But as Korolev was honored with a hero's recognition and burial in Red Square, his newest creation—an even larger "third generation" spacecraft, *Soyus 1* (pronounced Suh-Yoosh)—was being fashioned in Western Russia. An obscure Soviet technical journal even ran a picture of an "experimental spacecraft," later believed to have been Soyus. The long cylindrical spacecraft with its row of round ports giving it the appearance of a commuter train, was not taken seriously by the few westerners who chanced to see its likeness. An exception was Joseph Zygielbaum, a California Russia-watcher, who said to himself as he clipped the obscure photograph, "That may be Korolev's next baby."

The first sign that a new spacecraft was being tested in Russia was the appearance of one or more so-called Cosmos satellites with low perigees and brief flights. The assumption was that this was unmanned rehearsal flight. Using this technique, players of the space game in Washington could belatedly see preflight evidence of the Gagarin launch in the five unmanned, low perigees flights that just preceded it. Each Voskhod was also foreshadowed by a single rehearsal flight. In late 1966, the Washington orbital oracles noted the eccentric *Cosmos 133* which had a low orbit and remained up for only two days. A few weeks later *Cosmos 140* repeated these characteristics, and several other Cosmos satellites soon exhibited strangely eccentric behavior.

At about the same time Dr. Charles Sheldon of the President's Space Council observed that touring cosmonauts had mentioned in three different countries that a future spaceship would carry "more than five passengers." In fact, by early 1967 a surprising number of Russians were talking ambitiously of flights to the moon and of advanced spacecraft then under development on the ground. The Russian temptation to talk in advance, as the U.S. habitually did, was slowly getting the best of those stern Soviet disciplinarians who assuredly still argued for a "shoot now—talk later" policy.

The first verbal hints that something was up were made in the Moscow area beginning about mid-April of 1967. There are approximately two dozen American correspondents permanently stationed in Moscow. To most of these, especially the new ones without firm contacts of their own, the reaction of three key journalists to whatever rumors happened to be current in any given week is considered as reliable a barometer as any other in Russia. These three are Henry Shapiro of UPI, the "old man" of foreign correspondents in Moscow; Edmund Stevens, another old-timer of very gentle manners who drives the only red Mercedes in Russia (it was Ed Stevens who had headed the Moscow Bureau of *Time* before *Time* was expelled from the U.S.S.R. for articles unfavorable to the government); and, perhaps the most important indicator, an inscrutable, personable, wealthy Russian named Victor Louis. I had once been among the guests of Victor and his charming English wife, Jennifer, as they entertained most of the influential members of the Moscow press corps at their handsome three-story *dacha* at Peredelkino—the picturesque village some twenty miles from Moscow, where Boris Pasternak lived and wrote. Victor has written for a number of American publications but his regular job, the one that sustains his accreditation as a journalist, is his assignment as Moscow correspondent for the *London Evening News*. From mid-April on information came to Shapiro, Stevens, and Louis that the Soviets were planning something special; the other correspondents, and the world at large, reacted accordingly.

The most consistent prognosis was that a new spacecraft was about to be launched and that, after rendezvous and docking, as many as five men would transfer from one craft to the other before the spacecraft returned separately to earth. The most interesting aspect of the rumor, as a rumor, was its redundancy. Why as many as five men? Rendezvous and docking, of course, was still considered essential to a manned voyage to the moon, but the transfer of any more than two men appeared superfluous to that mission. The significance rested in the fact that a five-man space bus was obviously somewhat beside the point as far as the moon was concerned; its appearance at this point in the schedule would indicate that the Russians were allowing tangential missions and had, therefore, abandoned any possible hope of beating the Americans to the moon in favor of a more mundane spectacular.

The world learned of the existence of *Soyus 1* and the name of its single passenger, Colonel Vladimir Komarov, on April 23, a little over a month after Komarov's fortieth birthday. Tass, contrary to earlier practice, gave only a few details: *Soyus 1* (*Union 1*) had been launched from Baikonur at 3:35 A.M. with Komarov aboard and alone. By 10:00 A.M. it had completed its fifth orbit and "the flight was proceeding according to plan." In mid-afternoon, Tass indicated Komarov would be out of touch from 1:30 P.M. until 9:30 P.M. and would rest during that period. (Due to a lack of worldwide Soviet tracking facilities, the Russians, unlike the U.S., were heavily penalized with a long radio blackout period during each orbit.) The next report, which came after midnight, stated that *Soyus 1* had completed 13 orbits. The final flight report came at 6:12 A.M. the morning of the 24th. "According to the report of the pilot-cosmonaut, Komarov, and telemetered information, he is in good health and feels well. The ship's systems are functioning normally."

There were three surprises so far inherent in the Soviet actions. First, why such terse announcements and so little of the usual fanfare over a manned launch? Second, why had the manned inclination to the Equator been shifted from 65° to 52° (a fact we quickly observed with delicate instruments), an inclination more favorable for an eventual lunar-launch trajectory? And, third, why had Komarov been launched alone in what was presumably a brand-new spacecraft?

My own reactions at this stage were mixed. I had been most impressed with Komarov when I talked with him in Russia. His personality radiated warmth and he had what is known as "high visibility" even when he moved in a crowd. At the conclusion of our talks, when I presented him with my Gemini tie clasp, his bright brown eyes showed an instant and almost boyish delight. Then, with a natural politeness, he removed a gold and ruby pin from his colonel's uniform and, with a big grin pinned it on my coat. As he did so, his smile was as guileless, as sparkling, and as contagious as that of my friend and neighbor in Seabrook, Texas, John Glenn.

The cosmonauts themselves were also impressed with Komarov. There were reports that his advice was frequently sought by his younger colleagues. In the Soviet documentary film on Alexei Leonov's walk in space, Komarov and Gagarin were the only cosmonauts who were intimately associated with and who escorted Leonov and Belyayev to their waiting rocket. Tass had indicated that Komarov's selection for a second space flight was based on "his knowledge of engineering, terrific capacity for work, and great doggedness."

I was pleased that Vladimir Komarov now held the honor of being the first cosmonaut in the Soviet Union to be assigned to two space flights. I felt the selection of a man already experienced in the vital functions of a space-crew commander was significantly important for the complex nature of the mission now underway.

The next electrifying announcement from Tass did three things: it vanquished, as the first reported tragedy in space, a highly personable, capable, and valiant pilot; it ended whatever immediate hopes the Soviets had for manned rendezvous, docking, and crew exchange; and it stopped the Soviet manned space-flight program in its tracks, much as the U.S. program had been stopped three months earlier when a fire in the Apollo spacecraft killed Gus Grissom, Ed White, and Roger Chafee on the pad.

The initial delayed announcement of the tragedy created about as many questions as it answered. In essence, it stated that Komarov had been killed when he attempted to enter the earth's atmosphere after his eighteenth orbit, due to the fact that "when the main cupola of the parachute opened at an altitude of seven kilometers [4.34 miles] the straps of the parachute . . . got twisted and the spaceship descended at great speed which resulted in Komarov's death."

The immediate reaction was a sort of numbness, that space, like the sea and, after that, the air, had now begun to exact one of the inescapable penalties of human penetration. There had been many foreshadowings of such an accident. John Glenn, for instance, had a deep personal concern that a sudden accident in the U.S. program would cause undue reactions from an unprepared public. He had gone out of his way many times, both publicly and privately to indicate that tragic accidents in the space program were not only probable but quite possible. Both Ed White and Gus Grissom pointedly mentioned the possibility of accidents several weeks before their last scheduled flight. More specifically, just two months before the Apollo accident, Dr. Eugene B. Konecci of the National Aeronautics and Space Council, remarked to a Boston audience, "We intuitively know that some day, maybe in the not too distant future, we or the Soviets will suffer a manned space catastrophe."

Now, after 22 manned excursions, involving 35 astronauts and cosmonauts who had collectively spent 2,500 hours beyond the confines of the atmosphere, space had preempted its toll.

The confused response, on both sides of the titanium curtain, at first overshadowed the very few details available on the how and the why. Walter Cronkite called from New York to discuss the accident, to ask about the sort of person Komarov was, and whether or not I knew anyone in Russia I could call on for facts or reactions before his evening CBS News program. The only Russian telephone number I had was that of Victor Louis. I put in a radio telephone call to Victor and, expecting it to take at least the average of about four hours, I left on an errand. I had been gone less than five minutes when the phone rang and my wife answered it. The connection was very poor, but she understood the person talking was Victor's wife, Jennifer, whom she had never met. Two very confused women shouted to make themselves understood.

"What is the reaction?" my wife, Helene, asked. "Are people very sad?"

"Of course we're very sad," Jennifer's crisp English accent responded. "I've been crying all morning. I am alone in the country and heard it on the radio. It's a terrible thing to happen."

After the wire services had called for an interview, Russian expert Joe Zygielbaum called Houston from California to say he had just intercepted a Radio Moscow broadcast on his short-wave set. He told me the facts, essentially what Tass had already announced. Later, I innocently suggested to UPI that they talk to Zygielbaum. What resulted was a thoroughly confused story that was, nevertheless, circulated worldwide. The account had the Russians encountering all sorts of difficulties throughout the flight well in advance of the previously announced difficulty with the parachute. As it turned out later, the erroneous story was due to a combination of a faulty connection and a misunderstanding. Zygielbaum said the new information in the story came from an area of conversation that was his own speculation, not the facts as stated by Radio Moscow.

Other news media reported errors and rumors and reported the wild surmises that characterize nearly every reaction to a major Russian launch, and more specifically, to what the Soviets announce about the event. The *Washington Star*, for instance, ran a story which leaned heavily on rumors that Komarov was dead before his attempt at re-entry. The *Star* (and later, *Time*) in what can only be described as an astounding lapse in editorial judgment, even reported that Komarov may have "had a premonition of his fate," which is the way *Time*—in the obvious absence of veteran science editor, Jack Leonard—put it. The rumor was that just before Komarov entered his spacecraft, he handed a copy of a biography of Joan of Arc to a Russian reporter named Sergei Borzenko. The book, according to the rumor, contained the following marked passage: "She bade her farewells and continued gazing at the clear blue sky until the final second when the black smoke blotted out that sky forever."

This grand operatic gesture would be roughly comparable to astronaut Wally Schirra solemnly handing—just before entering Apollo—the following quotation of the late Henry Luce to Ed Diamond of *Newsweek*: "To see the world, to see life, to eyewitness great events. . . ."

The number of persons who apparently support themselves by propounding a theme opposite from that stated by the Soviet spokesmen on space topics is astonishing. No one but a fool would believe everything the Russians stated, but our one hundred and eighty degree skepticism in this regard has become so predictably automatic, and so cheaply fashionable, that it is almost a way of life. And like most fashion, it is something put before the vision that obscures the truth.

Another byproduct of the tragic flight of *Soyus 1*, was the support the Soviet admission of the tragedy indirectly gave to those who had long disbelieved assertions that Soviet security habitually concealed deaths in space. Yet, even so, the Komarov tragedy put some of the "concealed death" exponents discussed earlier back in business. According to UPI, a Stanford University employee named Julius Epstein said American authorities know of and conceal the names of 12 cosmonauts who have been lost in Russian space-flight mishaps. Besides the name of purported cosmonaut, Pioir Dolgow, Epstein mentions such curiously spelled names as Serenty Shiborin, Wassilievitch Zowodovsky, Alexei Belokonev, Ivan Kascheur, Alexis Graizev, and Jennady Michailov.

"Now is the time," Epstein is quoted as saying, "for the [U.S.] government to make the deaths public for the sake of accurate history."

Epstein, and others like him, is appealing to the same element in U.S. life that feels and perhaps wants to feel and must feel that, for instance, the U.S. Air Force or the government is concealing the existence of flying saucers from the American people.

Other stories of speculation, exploitation, or refutation

suggested that Komarov's spacecraft was tumbling badly in orbit long before his re-entry difficulties and that his difficulty had been apparent as early as his fifth orbit. Dr. Charles Sheldon, in his excellent "Review of the Soviet Space Program," pointed out it was the Western—not the Russian—press that enhanced the original launch story with the information that *Soyus 1* was put up by "the most powerful rocket yet." Sheldon points out the elementary flaw of this kind of dramatic embellishment: that neither the U.S.S.R. nor the U.S. would first fire "the most powerful rocket yet" with a man on top of it. One or more flights are always required to man-rate any new rocket.

What actually caused the tragedy was probably the parachute shroud-line difficulty the initial announcement referred to. This type of accident, for one thing, reflects the law of averages in any experimental parachute-braked test vehicle, especially an exceedingly heavy one. In fact, a tangled parachute harness was one of the things that had worried engineers working on Apollo descent systems at Houston's Manned Spacecraft Center. In all likelihood, *Soyus*, like *Voskhod* and *Apollo*, was to have used three main parachutes, each of which, after drogue deployment, was to be first extended by a small pilot chute. If one of the three pilot chutes is slightly late in deploying, it is quite conceivable that its nylon lines could cross the lines of another pilot chute and, in deploying, hopelessly bundle, tangle, and possibly sever lines and shrouds.

The other question, what happened to the rendezvous experiment, is probably explained by the fact that Komarov attempted re-entry on his eighteenth orbit, this possibly could have occurred after trouble had developed on the ground with the second spacecraft or carrier rocket. Calculations show that the best orbit for a rendezvous would have been the sixteenth when the ground track of *Soyus 1* again passed near the Tyuratam launch site. When the second, or rendezvous, part of the mission became impossible—conceivably at a time close to that of the planned second launch—there was no point in his staying up. In such a contingency, Komarov would need to alter his schedule and make detailed preparations to land;

the eighteenth would have been a logical orbit in which to try to descend.

The low key of the original Soviet announcements of launch was possibly due to the fact that they wanted to give maximum exposure to subsequent announcements of a "union" in space.

There is one area of pure speculation on the cause of the parachute difficulty that might be relevant: that is the possibility that the new transfer tunnel, or air lock, either caused unpredictable turbulence which interfered with normal parachute deployment, or that the air-lock protuberance itself fouled the parachute system.

In this connection, it should be remembered that *Voskhod 2* contained, for the first time, an air lock, at least a portion of which protruded from the spacecraft. It is highly unlikely that in *Soyus* the Soviets would have gone to the trouble of testing out a decidedly different and, therefore, risky new air lock or transfer tunnel. While documentary film clearly shows the *Voskhod 2* tunnel to occupy a relatively small portion of the cockpit area—roughly the amount taken up by the missing third seat—the film also establishes that the tunnel is, at least, as long as a man at the time it is used. An unknown—and never clearly revealed portion of it protrudes from the outside of the spacecraft. In ground pictures the Soviets released of the sheathed *Voskhod 2* (with protective fairing in place), no large protuberance is shown, although we can perceive a sort of opaque, round blister that was not present on *Voskhod 1*. Scale projection shows that this blister could protect, or conceal, a protuberance of certainly less than two feet. This would suggest a tunnel of only about four feet in length, certainly one incapable of accommodating Colonel Leonov. The rough drawing of the spacecraft made by Leonov after his flight, shows a fairly large external protuberance, certainly a longer one than can be accounted for by ground pictures of *Voskhod 2*.

The entire air-lock assembly, therefore, must either telescope or accordion. But even in its most aerodynamically clean configuration, some portion of it still protrudes beyond the clean lines of the spacecraft. Whether it remained locked in the "in" position or was forced into the "extended" position by entry factors—such as G forces or unplanned centrifugal forces—it undoubtedly would add two risk factors previously present only once in a Soviet spacecraft: first, a protuberance that could transfer aerodynamic turbulence, or thermal heating, to the spacecraft area which deployed the parachute system and second, less likely, a protuberance that could physically foul the deployment system.

The arrangements for Colonel Komarov's state burial were themselves a reminder of the earlier state burial of Sergei Korolev, the Chief Spacecraft Designer. Was there any connection between Korolev's first absence from a manned flight and Komarov's fatal difficulties? Probably not a man in Moscow knew the answer to that question; if, indeed, it could be answered. Komarov's ashes were placed in a gray metal urn displayed on the flower-decked bier on the second floor of a building known as the House of the Soviet Army. Lines were two-miles long as an estimated 100,000 persons filed past. Later, the ashes were placed in a niche in the Kremlin wall, the same wall that held the ashes of Sergei Korolev.

A month after Komarov's death Yuri Gagarin gave his version of the accident to the newspaper *Komsomelskaya Pravda*. The orbital flight, Gagarin said, "passed absolutely normally and did not cause even a shadow of alarm. . . . At the finish everything was excellent until the moment the parachute system had to go into operation."

The loss of Vladimir Komarov was a consequence of that curiosity in man that has always been instinctive and evolutionary and which, in the majority of its applications, ultimately enriches life on this planet.

The death of Komarov, the first loss that could be directly attributable to space flight, brought to three the total number of Soviet cosmonauts killed either in training or space flights; one had died earlier in a parachute accident and another had been killed in a training flight in an airplane.

The United States had suffered more grievously. By the spring of 1968 five astronauts—Ted Freeman, Charles Bassett, Elliot See, Clifton Williams, and Robert Lawrence had been killed in aircraft accidents. Lawrence was a Negro astronaut in training in the U.S. Air Force's Manned Orbital Laboratory program. Three—Ed White, Gus Grissom, and Roger Chaffee—had died during a ground check out of the Apollo spacecraft, and a ninth astronaut, Edward Givens, had been killed in an automobile accident.

On March 27, with dramatic and inexplicable suddenness, a distinguished new name was added to the list. Yuri Gagarin, according to an official Moscow report, had been killed during an airplane flight, along with his companion and backup cosmonaut, Colonel Vladimir S. Seryogin. According to reports received in the U.S., Victor Louis had talked to friends of Gagarin who said the two pilots crashed in a MIG 15 jet that had been converted to carry two. The crash was reported to have occurred about thirty miles east of Moscow near the town of Vladimir. Few other details were given.

For the third time, Moscow prepared for the ceremonial burial of a space hero in Red Square. The ashes of both Gagarin and Seryogin were placed in the red brick wall of honor near Lenin's tomb.

As thousands of Russians, including top government officials, massed in Red Square, Cosmonaut Andrian Nikolayev spoke: "Our sorrow is deep and painful and incurable. . . . He was rightly called the first man of the universe. . . . Nothing will stop us on our way to conquer outer space. We are ready."

Nearby, Gagarin's wife Valentina, embraced and kissed her husband's picture as Nikolayev's own wife, the former Valentina Tereshkova, also wept. In an unusual gesture, auto horns, factory whistles, and bells all sounded at once, and then were followed by a moment of silence.

The loss of Yuri Gagarin cost the Soviet Union much more than an inviolate symbol of space conquerer and state hero. In the six years since his pioneering flight, he had assumed duties far transcending those of a space pilot. First of all, as we have seen, he served as tutor and coach to other cosmonauts. Secondly, he served as official deputy to Lt. General Nikolai Kamanin, chief of the Soviet space program commission. And finally, and perhaps most significantly, he was

A military rocket said to be capable of orbiting "global" bombs; on display during May Day parade in Red Square in 1967.

The Moscow obelisk in honour of Soviet space exploration. A statue of Konstantin Tsiolkovsky stands at the foot of the monument. The granite obelisk is coated with non-corrosive titanium, a space-age metal widely used in the spacecraft of both the East and the West.

in charge of all projects involving lunar exploration. A number of observers, including some in Russia, expected him to lead the first Soviet cosmonaut team assigned to land on the moon. Since he was only 34 at the time of his death, he would have possessed that fortunate blend of experience and youth required to command a voyage to the moon. His presence as spaceship commander would have given confidence to fellow crew members, to mission-control personnel, and to the entire Soviet Union.

But now it is not to be. In Gagarin and in Komarov, General Kamanin and the Soviet Union have lost two of the most valuable and qualified of its space pilots. No Soviet pioneering flight to the moon could possibly proceed without recalling their unique qualifications for assuming a leading role in such an undertaking.

K

17
Soyus

IN SEPTEMBER 1968 the Russians achieved a dramatic and demonstrable success with Zond 5, an unmanned spacecraft which circled the moon and returned to a safe landing on earth. It was immediately widely predicted that the next Soviet launch would attempt to send one or more men around the moon, but there were two factors which made such an attempt out of phase, or premature, in Soviet terms.

In the first place, Russian fears, which I had first heard at the Soviet Academy of Sciences in Moscow, that the moon's surface and vicinity might be dangerously high in radiation had not been entirely eliminated, at least in the minds of some high placed Soviet scientists. A far more logical next circumlunar flight would be one containing a dog or other life forms to test possible bad effects of "secondary" radiation emanating from the moon's surface and the effects of the earth's comet-shaped radiation tail on life near the moon (elements of the Soviet press had severely criticized U.S. plans to send men to the moon's vicinity without first checking the route with animals). In the second place, although the Russians never repeated a specific space flight if it were successful, they invariably repeated unsuccessful flights, as shown especially, by the long string of virtually identical missions they had originally launched at the planets and at the surface of the moon. The fact still to be surmounted was that the first and last flight involving their new spaceship, Soyus 1, had killed its pilot, Vladimir Komarov, during an attempted return to earth. In addition, the unusually well founded rumors associated with that tragic flight, had suggested that some sort of rendezvous experiment had also been planned with Soyus, which means "union" in Russian. The paramount fact was that Soyus had become the first spacecraft to fail its baptism in space. In the dogged logic of Soviet spacemen, the errors had to be corrected and the flight had to be repeated and consummated.

The Soyus tragedy, like the U.S. Apollo disaster, had caused well over a year's delay in the manned space flight program. So serious was it considered in the Soviet Union that Russian newspapers took the unusual measure of coming very close to making Western style demands on Soviet officials for a full investigation. In the absence of the commanding and unimpeachable Chief Designer, Sergei Pavlovich Korolev, whose word alone would have dissipated most doubts and possible moods of recriminations, the 18 months devoted to Soyus modification and rectification must have been frenetic ones.

For well over a year the Soviets gave no hint as to what was wrong with Soyus, other than that the parachute had failed. Nor did any Russian spokesman elaborate on any uncompleted portion of Soyus' maiden flight or on such rumors as that reported by UPI in Moscow that cosmonaut Valery Bykovsky —a veteran of the earlier dual "Romanov and Juliet" flight— had been waiting on the ground to link up with Komarov in Soyus 1.

In late 1968, long time Soviet space spokesman, Academician Leonid Sedov, visited Rice University in Houston, Texas to present a lecture. Preceding his talk, I joined him at Rice to ask him about several aspects of the Soviet space program, including the Soyus disaster.

"The reason for the Soyus tragedy," Sedov said through a fellow scientist interpreter, G. G. Chernyi, "was the malfunction of the parachute system. Before Soyus 1, we had several tests of the same system and the same ship without a cosmonaut aboard; all tests were successful. So it was a very much unexpected difficulty."

Sedov furnished no other details, other than to admit that corrective work was extensive. "The difficulty," he said, "was not enough to raise serious questions about the continuation of the Soyus program, but a large amount of modification work was done."

During the week of October 19, 1968, as U.S. astronauts Walter Schirra, Donn Eisele and Walter Cunningham concluded their 11 day mission in Apollo 7, the Soviets launched into orbit Cosmos numbers 248 and 249 which U.S. trackers quickly observed to be in the same plane. Although these launches made no headlines, they were a reasonable indicator to some U.S. intelligence sources that the Russians were ready to repeat at least a portion of the Soyus 1 mission.

The actual beginning of the next manned Soviet flight was a highly secret affair. On Friday October 25, two widely scattered but interesting events took place: Communist party leader Leonid Brezhnev told a meeting of the Communist Youth League in Moscow that Russia would have "direct acquaintance with the moon," and in the U.S., trackers saw a new Soviet object in orbit. Although Russian sources remained silent, this, as Brezhnev most certainly knew, was actually the long awaited Soyus 2—presumably unmanned. Its unannounced apogee was 139.1, its perigee 114.8 and its inclination from the equator, 51.7 degrees.

The first word from Russia came the next day when a second object appeared in orbit with an announced apogee of 135 miles and a perigee of 123 miles, close to the plane of Soyus 2 and virtually identical to the characteristics of Soyus 1. This was Soyus 3, the Russians announced. Without even identifying Soyus 2, they stated that Soyus 3 was piloted by cosmonaut Colonel Georgi Timofeyevich Beregovoy.

Now Tass, Pravda, Radio Moscow and other outlets in the U.S.S.R. began a somewhat constrained report of flight details and color. Cosmonaut Beregovoy at 47 was not only the oldest man yet to fly in space, he was also older than any of the 52 U.S. astronauts (Wally Schirra was 45 when he announced that his flight in Apollo 7 would be his last, partly because of his age). Beregovoy, at 47, was old enough to have been a decorated veteran of Russian squadrons which opposed the German Luftwaffe in World War II, a 30 year veteran of the Red Air Force and the father of a university age son, Victor. Unlike the names of most cosmonauts, which are of ethnic origin, Beregovoy is a common Ukrainian name, especially around the Poltava region where the young metallurgical worker grew up. He entered the Red Air Force School at Lugansk in 1938 and in the final year of the war in Europe, was made a Hero of the Soviet Union. After the war, he won the rare title "Merited Test Pilot of the U.S.S.R.". The studious looking, brunette colonel was a late comer to the Soviet cosmonaut program; he did not even apply for cosmonaut training until 1964, after the entire one man Vostok flight series had been completed.

Soviet news media indicated enigmatically that Beregovoy on his very first orbit automatically maneuvered within about 200 yards of his target vehicle, Soyus 2, in a maneuver the Russians call the *priblizhenye*, or approach. Then he was reported to have manually maneuvered even closer, although his closest point was never precisely stated. The next day, Sunday, he rendezvoused with Soyus 2 and "performed extensive maneuvers" for 90 minutes. To the surprise of many observers, no announcement was made, then or later, that he had actually docked, although Pravda had previously stated that one purpose of his mission was "to perfect docking techniques in orbit." One announcement did say that docking maneuvers were "fully carried out," but nothing more definitive was forthcoming. At one point on Sunday when Beregovoy appeared on satellite-to-earth television, the ground control commission was heard to say, "We have evaluated your day's work and find that you have carried out the program excellently."

To which Beregovoy merely replied, "*Ponyal, spasibo* (Roger, thank you)."

Was it really possible that a nation which had twice accomplished automatic docking between unmanned satellites and which had twice before brought two manned spacecraft within 3 miles of each other, was it possible that such a nation would then schedule a mission that called only for a close approach of one spacecraft to another, that did not call for complete docking?

When I saw Leonid Sedov just after the Soyus 3 flight, I asked him this question directly. "Twice," I said, "your cosmonauts have come within 3 miles of each other. And now, if we are to believe the Soyus 3 reports, your cosmonaut has again failed to join up in orbit. Why? You are like lovers who kiss and tease but nothing else. Why do you not consummate? Why do you not go all the way in?"

Cagey, shrewd old Sedov, who was making public comments on the Soviet space program as long ago as 1955, looked tolerantly at me over his spectacles. His eyes were creased in an affable grin. He appeared the soul of inscrutable innocence and self-confidence.

"But I was out of the country when it happened!" he replied. "We have done automatic docking and that is more difficult."

As Beregovoy continued to orbit, a few extra details came through. No exact description of Soyus was given except that it was large and had two compartments, or rooms. The combination "bedroom" and scientific compartment contained a desk and locker; the other room was the control room which Russian television viewers saw briefly when Beregovoy donned in coveralls panned his spacious quarters with his on-board TV camera. Some viewers thought they detected enough room for at least two more cosmonauts.

There appeared to be two control handles—one, apparently, to ignite and control the engine which accelerated or braked Soyus, the other to control Soyus' rotation around its center of mass.

At one point Beregovoy put on a special harness which contained temporary medical sensors which supplemented the permanent ones he wore against his body. Academician Vasili V. Parin, with whom I talked in Moscow, later wrote that not only could Beregovoy's sensors "send coded signals to earth indicating some health disorder but also could inform the (future) crew of hazards, and even switch on emergency safety units." There is the possibility, of course, that such devices could trigger an automatic landing should a cosmonaut become incapacitated in space.

While Beregovoy was in his second day of orbit, the Soviets unaccountably brought Soyus 2 home to a successful parachute landing, and on Tuesday, October 29, Beregovoy himself came to earth in Soyus 3. He reportedly fired his retro-rocket in a new manner, leaving it on for over a minute so that it was still burning and adjustable even after entering the atmosphere. He entered earth's atmosphere on his 61st orbit after 96 busy hours in space. The fact that the "bedroom" compartment was jettisoned indicated to at least one Russian watcher, Joseph Zygielbaum, that the Russians were shifting over to a modular type of spacecraft which utilized an advanced type of sliding door between sections. During the snowdrift landing near a Kazakh village, he successfully used both the new Soyus parachute system and Voskhod type landing rockets. He emerged in the same street clothes he was seen to wear on television when he entered his spacecraft. The first things he asked for were a fur coat and a bowl of soup.

In hailing his flight, the Russians spoke of the ability to maneuver their new cosmoplane "like an automobile" instead of following tracks "like a railroad train." Celebrations in Moscow were as exuberant as ever. (Beregovoy was also promoted to General) but in the U.S. the feeling seemed to be that the Soviets were making a great deal of noise over what the rest of the world could only regard as an incomplete or aborted docking attempt. Unless, of course, the Russians were confidently lifting vodka glasses to celebrate a necessary prerequisite to a much more ambitious flight already standing in the wings.

Certain preparations pointed to high drama. For one thing, television and radio crews were getting their own needed brand of rehearsal. For the first time a reasonably complete documentary film of the launch was shown nationwide during the second day of the flight of Soyus 3. Significantly perhaps, Soviet radio announcers at the Baikonur cosmodrome were able to say that other spaceships were on the pads. As one of them put it, "strenuous work is underway."

CHRONOLOGY OF SOVIET SPACE FLIGHTS*

Name	International Designation	Date[1]	Payload Weight (pounds)	Period (minutes)	Perigee (miles)	Apogee (miles)	Inclination[2]	Status
Sputnik 1	1957 A 2	10-4-57	184	96.2	141	588	65.1	Decayed 1-4-58: first artificial satellite, transmitted 21 days
Sputnik 2	1957 B 1	11-3-57	1,120	103.7	140	1,038	65.3	Decayed 4-14-58: carried dog Laika, transmitted 7 days
Sputnik 3	1958 △ 2	5-15-58	2,925	105.8	140	1,168	65.2	Decayed 4-6-60: variety of scientific data returned up to decay
Luna 1	1958 M 1	1-2-59	797	450 days	.9766 AU[3]	1.314 AU	0.10	In solar orbit: lunar probe, passed within 3,728 mi. of moon
Luna 2	1959 Ξ 1	9-12-59	858	Flight time: 33.5 hours				Impacted on moon: first probe to hit the moon

CHRONOLOGY OF SOVIET SPACE FLIGHTS

Name	Designation	Launch Date						Remarks
Luna 3	1959 Θ 1	10-4-59	614	16.2 days	25,257	291,439	76.8	Decayed 4-20-60: photographed moon's far side for 40 min.
Sputnik 4	1960 E 1	5-15-60	10,008	91.3	194	229	65.0	Decayed 9-5-62: Vostok prototype, recovery failed 5-19-60 as cabin went into higher orbit, cabin decayed 10-15-65
Sputnik 5	1960 Λ 1	8-19-60	10,120	90.7	190	211	64.9	Re-entered 8-20-60: dogs Belka and Strelka recovered on orbit 18
None	None	10-10-60						Mars probe failed; announced by U.S. in 1962

* Reprinted by permission of *TRW Space Log*, published by TRW Systems Group, Redondo Beach, Calif. Since some assumptions were made in correlating this data, complete accuracy cannot be assured.
[1] Launch date from local time at launch site.
[2] Degrees from the equator.
[3] Astronomical Units; approx. 93 million miles.

CHRONOLOGY OF SOVIET SPACE FLIGHTS

Name	International Designation	Date	Payload Weight	Period	Perigee	Apogee	Inclination	Status
None	None	10-14-60						Mars probe failed: announced by U.S. in 1962
Sputnik 6	1960 P 1	12-1-60	10,060	88.6	116	165	65.0	Decayed 12-2-60: recovery attempt failed, canine cabin lost
Sputnik 7	1961 B 1	2-4-61	14,292	89.8	139	204	65.0	Decayed 2-26-61: believed to be Venus probe abort
Venus 1	1961 Γ 1	2-12-61	1,419	300 days	.7183 AU	1.0190 AU	0.58	In solar orbit: Venus probe, radio contact lost at 4.7M mi.
Sputnik 8	1961 Γ 3		14,292	89.7	123	198	65.0	Decayed 2-25-61: launched Venus 1 from parking orbit

CHRONOLOGY OF SOVIET SPACE FLIGHTS   287

Sputnik 9	1961 ө 1	3-9-61	10,340	88.5	114	155	64.9	Re-entered 3-9-61: cabin with dog Chernushka recovered
Sputnik 10	1961 I 1	3-25-61	10,330	88.4	111	153	64.9	Re-entered 3-26-61: dog Zvezdochka recovered after 17 orbits
Vostok 1	1961 M 1	4-12-61	10,417	89.1	112	203	65.0	Re-entered 4-12-61: first manned space flight, cabin with Y. Gagarin recovered in U.S.S.R. after 1 orbit, 1.8 hrs.
Vostok 2	1961 T 1	8-6-61	10,430	88.6	111	160	64.9	Re-entered 8-7-61: G. Titov landed after 17 orbits, 25.3 hrs.
Cosmos 1	1962 ө 1	3-16-62		96.4	135	609	49.0	Decayed 5-25-62: numerous scientific objectives announced as Cosmos research satellite series initiated

CHRONOLOGY OF SOVIET SPACE FLIGHTS

Name	International Designation	Date	Payload Weight	Period	Perigee	Apogee	Inclination	Status
Cosmos 2	1962 I 1	4-6-62		102.0	131	960	49.0	Decayed 8-19-63: unannounced payload
Cosmos 3	1962 N 1	4-24-62		93.8	142	447	49.0	Decayed 10-17-62: unannounced payload
Cosmos 4	1962 Ξ 1	4-26-62		90.6	184	205	65.0	Re-entered 4-29-62: first and only announced Cosmos recovery
Cosmos 5	1962 Y 1	5-28-62		102.8	126	994	49.1	Decayed 5-2-63: monitored Starfish artificial radiation
Cosmos 6	1962 A△ 1	6-30-62		90.1	168	221	48.9	Decayed 8-8-62: unannounced payload
Cosmos 7	1962 AI 1	7-28-62		90.1	130	229	65.0	Re-entered or decayed 8-1-62: unannounced payload

CHRONOLOGY OF SOVIET SPACE FLIGHTS 289

Vostok 3	1962 AM 1	8-11-62	10,428	88.3	114	156	65.0	Re-entered 8-15-62: A. Nikolayev landed by parachute after 64 orbits, 94.4 hrs.: part of first Soviet "group" flight
Vostok 4	1962 AN 1	8-12-62	10,428	88.4	112	158	65.0	Re-entered 8-15-62: P. Popovich landed by parachute after 48 orbits, 71.0 hrs.; came within 3.1 mi. of Vostok 3 on orbit 1
Cosmos 8	1962 AΞ 1	8-18-62		92.9	159	375	49.0	Decayed 8-17-63: unannounced payload
None	1962 AII 1	8-25-62		88.7	107	157	64.9	Decayed 8-28-62: probable Venus probe failure
None	1962 AT 1	9-1-62						Decayed 9-6-62: probable Venus probe failure

CHRONOLOGY OF SOVIET SPACE FLIGHTS

Name	International Designation	Date	Payload Weight	Period	Perigee	Apogee	Inclination	Status
None	1962 AΦ 1	9-12-62						Decayed 9-14-62: probable Venus probe failure
Cosmos 9	1962 AΩ 1	9-27-62		90.0	187	220	65.0	Re-entered or decayed 10-1-62: unannounced payload
Cosmos 10	1962 BZ 1	10-17-62		90.2	130	236	65.0	Re-entered or decayed 10-21-62: unannounced payload
Cosmos 11	1962 BΘ 1	10-20-62		96.1	152	572	49.0	Decayed 5-18-64: unannounced payload
None	1962 BI 1	10-24-62						Decayed 10-29-62: probable Mars probe failure

Mars 1	1962 BN 3	11-1-62	1,965	519 days	.9237 AU	1.604 AU	2.683	In solar orbit: Mars probe, lost earth lock at 65.9M mi.
None	1962 BΞ 1	11-4-62						Decayed 11-5-62: probable Mars probe failure
Cosmos 12	1962 BΩ 1	12-22-62		90.5	131	252	65.0	Re-entered or decayed 12-30-62: unannounced payload
None	1963 1A	1-4-63						Decayed 1-5-63: probable lunar probe failure
Cosmos 13	1963 6A	3-21-63		89.8	127	209	65.0	Re-entered or decayed 3-29-63: unannounced payload
Luna 4	1963 8B	4-2-63	3,135		55,800	434,000		In barycentric orbit: lunar probe, missed moon by 5,281 mi.
Cosmos 14	1963 10A	4-13-63		92.1	165	318	49.0	Decayed 8-29-63: unannounced payload

CHRONOLOGY OF SOVIET SPACE FLIGHTS

Name	International Designation	Date	Payload Weight	Period	Perigee	Apogee	Inclination	Status
Cosmos 15	1963 11A	4–22–63		89.8	107	231	65.0	Re-entered or decayed 4–27–63: unannounced payload
Cosmos 16	1963 12A	4–28–63		90.4	129	249	65.0	Re-entered or decayed 5–8–63: unannounced payload
Cosmos 17	1963 17A	5–22–63		94.8	162	490	49.0	Decayed 6-2-65: unannounced payload
Cosmos 18	1963 18A	5–24–63		89.4	130	187	65.0	Re-entered or decayed 6–2–63: unannounced payload
Vostok 5	1963 20A	6–14–63	10,428	88.3	109	138	65.0	Re-entered 6–19–63: V. Bykovsky landed by parachute after 81 orbits, 119.1 hrs.; part of second "group" flight

Vostok 6	1963 23A	6-16-63	10,428	88.4	114	145	65.0	Re-entered 6-19-63: V. Tereshkova landed by parachute after 48 orbits, 70.8 hrs.; passed within 3 mi. of Vostok 5
Cosmos 19	1963 33A	8-6-63		92.2	168	322	49.0	Decayed 3-30-64: unannounced payload
Cosmos 20	1963 40B	10-18-63		99	123	186	65.0	Re-entered or decayed 10-30-63: unannounced payload
Polyet 1	1963 43A	11-1-63		102.3	212	868	59.0	In orbit: first spacecraft with extensive maneuver capability
Cosmos 21	1963 44A	11-11-63		88.5	121	142	64.8	Re-entered or decayed 11-14-63: unannounced payload
Cosmos 22	1963 45A	11-16-63		90.3	127	245	64.9	Re-entered or decayed 11-22-63: unannounced payload

CHRONOLOGY OF SOVIET SPACE FLIGHTS

Name	International Designation	Date	Payload Weight	Period	Perigee	Apogee	Inclination	Status
Cosmos 23	1963 50A	12-13-63		92.9	149	381	49.0	Decayed 3-27-64: unannounced payload
Cosmos 24	1963 52A	12-19-63		90.5	131	254	65.0	Re-entered or decayed 12-28-63: unannounced payload
Electron 1	1964 6A	1-30-64		169	252	4,412	61	In orbit: to study inner Van Allen radiation belt
Electron 2	1964 6B			1,360	286	42,377	61	In orbit: to study outer belt; first dual Soviet launch
Cosmos 25	1964 10A	2-27-64		92.3	169	327	49	Decayed 11-21-64: unannounced payload
Cosmos 26	1964 13A	3-18-64		91.1	168	250	49	Decayed 9-28-64: unannounced payload

CHRONOLOGY OF SOVIET SPACE FLIGHTS 295

Cosmos 27	1964 14A	3–27–64	88.4	119	147	64.8	Re-entered or decayed 3–28–64: unannounced payload
Zond 1	1964 16D	4–2–64	Heliocentric Orbit				In solar orbit: Venus probe, failed to return planetary data
Cosmos 28	1964 17A	4–4–64	90.4	130	245	65	Re-entered or decayed 4–12–64: unannounced payload
Polyet 2	1964 19B	4–12–64	92.5	193	311	58.1	In orbit: carried out maneuvers during first day in orbit
Cosmos 29	1964 21A	4–25–64	89.5	127	192	65.1	Re-entered or decayed 5–2–64: unannounced payload
Cosmos 30	1964 23A	5–18–64	90.2	128	238	64.9	Re-entered or decayed 5–26–64: unannounced payload
Cosmos 31	1964 28A	6–6–64	91.6	142	316	49	Decayed 10–20–64: unannounced payload

CHRONOLOGY OF SOVIET SPACE FLIGHTS

Name	International Designation	Date	Payload Weight	Period	Perigee	Apogee	Inclination	Status
Cosmos 32	1964 29A	6-10-64		89.8	130	207	51.3	Re-entered or decayed 6-18-64: unannounced payload
Cosmos 33	1964 33A	6-23-64		89.4	130	182	65	Re-entered or decayed 7-1-64: unannounced payload
Cosmos 34	1964 34A	7-1-64		90	127	223	65.0	Re-entered or decayed 7-9-64: unannounced payload
Electron 3	1964 38A	7-11-64		168.1	251	4,365	61	In orbit: to monitor inner Van Allen belt radiation
Electron 4	1964 38B			1,313.8	379	41,078	60.1	In orbit: to simultaneously study outer belt, magnetosphere

CHRONOLOGY OF SOVIET SPACE FLIGHTS

Cosmos 35	1964 39A	7-15-64	89.2	135	166	51.3	Re-entered or decayed 7-23-64: unannounced payload
Cosmos 36	1964 42A	7-30-64	91.9	161	313	49	Decayed 2-28-65: unannounced payload
Cosmos 37	1964 44A	8-14-64	89.5	127	186	65	Re-entered or decayed 8-22-64: unannounced payload
Cosmos 38	1964 46A	8-18-64	95.2	130	544	56.2	Decayed 11-8-64: first Soviet triple payload launch
Cosmos 39	1964 46B		95.2	130	544	56.2	Decayed 11-17-64: unannounced payload
Cosmos 40	1964 46C		95.2	130	544	56.2	Decayed 11-18-64: unannounced payload
Cosmos 41	1964 49D	8-22-64	715	245	24,765	64	In orbit: orbit similar to those later used by U.S.S.R. comsats

CHRONOLOGY OF SOVIET SPACE FLIGHTS

Name	International Designation	Date	Payload Weight	Period	Perigee	Apogee	Inclination	Status
Cosmos 42	1964 50A	8-22-64		97.8	144	683	49	Decayed 12-19-65: first double Cosmos launch
Cosmos 43	1964 50C			97.8	144	683	49	Decayed 12-27-65: unannounced payload
Cosmos 44	1964 53A	8-28-64		99.5	384	534	65	In orbit: unannounced payload
Cosmos 45	1964 55A	9-13-64		90	128	203	64.9	Re-entered or decayed 9-18-64: unannounced payload
Cosmos 46	1964 59A	9-24-64		89.2	134	168	50.3	Re-entered or decayed 10-2-64: unannounced payload
Cosmos 47	1964 62A	10-6-64		90.0	110	257	64.8	Re-entered or decayed 10-7-64: unannounced payload

CHRONOLOGY OF SOVIET SPACE FLIGHTS 299

Voskhod 1	1964 65A	10-12-64	11,731	90.1	100	255	65	Re-entered 10-12-64: first three-man crew—V. Komarov, K. Feoktistov, B. Yegorov, landed after 16 orbits, 24.3 hrs.
Cosmos 48	1964 66A	10-14-64		89.4	122	177	65	Re-entered or decayed 10-20-64: unannounced payload
Cosmos 49	1964 69A	10-24-64		91.8	162	304	49.0	Decayed 8-21-65: unannounced payload
Cosmos 50	1964 70A	10-28-64		88.7	122	150	51.3	Decayed 12-5-64: unannounced payload
Zond 2	1964 78C	11-30-64			Heliocentric Orbit			In solar orbit: Mars probe, failed to return planetary data
Cosmos 51	1964 80A	12-10-64		92.5	164	344	48.8	Decayed 11-14-65: unannounced payload
Cosmos 52	1965 1A	1-11-65		89.5	127	189	65.0	Re-entered or decayed 1-19-65: unannounced payload

CHRONOLOGY OF SOVIET SPACE FLIGHTS

Name	International Designation	Date	Payload Weight	Period	Perigee	Apogee	Inclination	Status
Cosmos 53	1965 6A	1-30-65		98.7	141	741	48.8	In orbit: unannounced payload
Cosmos 54	1965 11A	2-21-65		106.2	174	1,153	56.1	In orbit: unannounced payload, second Soviet triple launch
Cosmos 55	1965 11B			106.2	174	1,153	56.1	In orbit: unannounced payload
Cosmos 56	1965 11C			106.2	174	1,153	56.1	In orbit: unannounced payload
Cosmos 57	1965 12A	2-22-65		91.1	109	318	64.8	Re-entered or decayed 2-22-65: exploded into 150+ pieces
Cosmos 58	1965 14A	2-26-65		96.8	361	409	65.0	In orbit: unannounced payload

CHRONOLOGY OF SOVIET SPACE FLIGHTS 301

Cosmos 59	1965 15A	3–7–65		89.8	130	211	65.0	Re-entered or decayed 3–15–65: unannounced payload
Cosmos 60	1965 18A	3–12–65		89.1	125	178	64.7	Re-entered or decayed 3–17–65: unannounced payload
Cosmos 61	1965 20A	3–15–65		105	170	1,141	56.0	In orbit: unannounced payload, third triple Cosmos launch
Cosmos 62	1965 20B			105	170	1,141	56.0	In orbit: unannounced payload
Cosmos 63	1965 20C			105	170	1,141	56.0	In orbit: unannounced payload
Voskhod 2	1965 22A	3–18–65	13,200	90.9	107	308	65	Re-entered 3–19–65: A. Leonov spent 10 min. outside spacecraft, landed with P. Belyayev after 17 orbits, 26.0 hrs.

CHRONOLOGY OF SOVIET SPACE FLIGHTS

Name	International Designation	Date	Payload Weight	Period	Perigee	Apogee	Inclination	Status
Cosmos 64	1965 25A	3-25-65		89.2	128	168	65	Re-entered or decayed 4-3-65: unannounced payload
Cosmos 65	1965 29A	4-17-65		89.8	130	213	65	Re-entered or decayed 4-25-65: unannounced payload
Molniya 1	1965 30A	4-23-65		708	309	24,470	65	In orbit: first Soviet comsat, period adjusted to 12.0 hrs.
Cosmos 66	1965 35A	5-7-65		89.3	122	181	65	Re-entered or decayed 5-15-65: unannounced payload
Luna 5	1965 36A	5-9-65	3,254					Impacted on moon: first U.S.S.R. soft-landing attempt failed

CHRONOLOGY OF SOVIET SPACE FLIGHTS 303

Name	Designation	Date	Weight	Period	Perigee	Apogee	Inclination	Remarks
Cosmos 67	1965 40A	5-25-65		89.9	129	217	51.8	Re-entered or decayed 6-2-65: unannounced payload
Luna 6	1965 44A	6-8-65	3,179	Heliocentric Orbit				In solar orbit: lunar soft-lander, missed moon by 100,000 mi.
Cosmos 68	1965 46A	6-15-65		89.8	127	208	65	Re-entered or decayed 6-23-65: unannounced payload
Cosmos 69	1965 49A	6-25-65		89.7	131	206	65	In orbit: unannounced payload
Cosmos 70	1965 52A	7-2-65		98.3	140	717	48.8	In orbit: unannounced payload
Cosmos 71	1965 53A	7-16-65		95.5	342	342	56.1	In orbit: unannounced payload, first Soviet five-satellite launch
Cosmos 72	1965 53B			95.5	342	342	56.1	In orbit: unannounced payload

CHRONOLOGY OF SOVIET SPACE FLIGHTS

Name	International Designation	Date	Payload Weight	Period	Perigee	Apogee	Inclination	Status
Cosmos 73	1965 53C			95.5	342	342	56.1	In orbit: unannounced payload
Cosmos 74	1965 53D			95.5	342	342	56.1	In orbit: unannounced payload
Cosmos 75	1965 53E			95.5	342	342	56.1	In orbit: unannounced payload
Proton 1	1965 54A	7-16-65	26,896	92.5	118	390	63.5	Decayed 10–11-65: physics "lab," heaviest Soviet payload yet
Zond 3	1965 56A	7-18-65		Heliocentric Orbit				In solar orbit: retransmitting photos taken during lunar flyby
Cosmos 76	1965 59A	7-23-65		92.2	162	330	48.8	In orbit: unannounced payload
Cosmos 77	1965 61A	8-3-65		89.3	124	181	51.8	Re-entered or decayed 8–11-65: unannounced payload

Cosmos 78	1965 66A	8-14-65	89.8	128	204	69	Re-entered or decayed 8-22-65: unannounced payload
Cosmos 79	1965 69A	8-25-65	90	131	223	64.9	Re-entered or decayed 9-2-65: unannounced payload
Cosmos 80	1965 70A	9-3-65	116.6	932	932	56	In orbit: unannounced payload
Cosmos 81	1965 70B		116.6	932	932	56	In orbit: unannounced payload
Cosmos 82	1965 70C		116.6	932	932	56	In orbit: unannounced payload
Cosmos 83	1965 70D		116.6	932	932	56	In orbit: unannounced payload
Cosmos 84	1965 70E		116.6	932	932	56	In orbit: unannounced payload
Cosmos 85	1965 71A	9-9-65	89.6	132	198	65	Re-entered or decayed 9-17-65: unannounced payload

CHRONOLOGY OF SOVIET SPACE FLIGHTS

Name	International Designation	Date	Payload Weight	Period	Perigee	Apogee	Inclination	Status
Cosmos 86	1965 73A	9-18-65		116.7	857	1,050	56	In orbit: unannounced payload, third Soviet five-payload launch
Cosmos 87	1965 73B			116.7	857	1,050	56	In orbit: unannounced payload
Cosmos 88	1965 73C			116.7	857	1,050	56	In orbit: unannounced payload
Cosmos 89	1965 73D			116.7	857	1,050	56	In orbit: unannounced payload
Cosmos 90	1965 73E			116.7	857	1,050	56	In orbit: unannounced payload
Cosmos 91	1965 75A	9-23-65		89.8	132	213	65	Re-entered or decayed 10-1-65: unannounced payload
Luna 7	1965 77A	10-4-65	3,318	Flight time 86.1 hours				Impacted on moon: retros fired early, soft landing failed

Molniya 1B	1965 80A	10–14–65	719	311	24,855	65	In orbit: second Soviet comsat, U.S.S.R.–France communication link
Cosmos 92	1965 83A	10–16–65	89.9	132	219	65	Re-entered or decayed 10–24–65: unannounced payload
Cosmos 93	1965 84A	10–19–65	91.7	137	324	48.4	In orbit: unannounced payload
Cosmos 94	1965 85A	10–28–65	89.3	131	182	65	Re-entered or decayed 11–5–65: unannounced payload
Proton 2	1965 87A	11–2–65	92.6	119	396	63.5	In orbit: heavyweight complex high-energy physics laboratory
Cosmos 95	1965 88A	11–4–65	91.7	129	324	48.4	In orbit: unannounced payload, broke up third day in orbit
Venus 2	1965 91A	11–12–65	2,123	Heliocentric Orbit			In solar orbit: planetary probe, on 3½-month flight to Venus

CHRONOLOGY OF SOVIET SPACE FLIGHTS

Name	International Designation	Date	Payload Weight	Period	Perigee	Apogee	Inclination	Status
Venus 3	1965 92A	11-16-65	2,116	Heliocentric Orbit				In solar orbit: planetary probe, backup to Venus 2
Cosmos 96	1965 94A	11-23-65		89.6	141	193	51.9	Decayed 12-9-65: suspected Venus probe left in parking orbit
Cosmos 97	1965 95A	11-26-65		108.3	137	1,305	49	In orbit: unannounced payload
Cosmos 98	1965 97A	11-27-65		92	134	354	65	Re-entered or decayed 12-5-65: unannounced payload
Luna 8	1965 99A	12-3-65	3,422					Impacted on moon: retros fired late, soft landing failed
Cosmos 99	1965 103A	12-10-65		89.6	124	199	65	Re-entered or decayed 12-18-65: unannounced payload

CHRONOLOGY OF SOVIET SPACE FLIGHTS

Name	Designation	Date					Remarks
Cosmos 100	1965 106A	12–17–65	97.7	404	404	65	In orbit: unannounced payload
Cosmos 101	1965 107A	12–21–65	92.4	162	342	49	In orbit: unannounced payload
Cosmos 102	1965 111A	12–28–65	89.2	135	173	65	Re-entered or decayed 1–13–66: unannounced payload
Cosmos 103	1965 112A	12–28–65	97	373	373	56	In orbit: unannounced payload, 52nd Cosmos orbited in 1965
Cosmos 104	1966 1A	1–7–66	90.2	127	249	65	Re-entered or decayed 1–15–66: unannounced payload
Cosmos 105	1966 3A	1–22–66	89.7	127	201	65	Re-entered or decayed 1–30–66: unannounced payload
Cosmos 106	1966 4A	1–26–66	92.8	180	350	48.4	In orbit: unannounced payload
Luna 9	1966 6A	1–31–66	220	Flight time: 79.0 hours			Landed on moon: returned photos of lunar surface for 3 days

CHRONOLOGY OF SOVIET SPACE FLIGHTS

Name	International Designation	Date	Payload Weight	Period	Perigee	Apogee	Inclination	Status
Cosmos 107	1966 10A	2-10-66		89.7	127	200	65	Re-entered or decayed 2-18-66: unannounced payload
Cosmos 108	1966 11A	2-11-66		95.3	141	537	48.9	In orbit: unannounced payload
Cosmos 109	1966 14A	2-19-66		89.5	130	192	65	Re-entered or decayed 2-27-66: unannounced payload
Cosmos 110	1966 15A	2-22-66		95.3	116	562	51.9	Re-entered 3-16-66: two dogs recovered after 330 revolutions
Cosmos 111	1966 17A	3-1-66		88.6	119	140	51.9	Decayed 3-3-66: suspected lunar probe failure

CHRONOLOGY OF SOVIET SPACE FLIGHTS

Cosmos 112	1966 21A	3-17-66	92.1	133	351	72	Re-entered or decayed 3-25-66: new inclination for U.S.S.R.
Cosmos 113	1966 23A	3-21-66	89.6	130	203	65	Re-entered or decayed 3-29-66: unannounced payload
Luna 10	1966 27A	3-31-66	540	178	217	72	In lunar orbit: lunar and circumlunar data until 5-30-66
Cosmos 114	1966 28A	4-6-66	90.1	130	232	73	Re-entered or decayed 4-14-66: unannounced payload
Cosmos 115	1966 33A	4-20-66	89.3	118	183	65	Re-entered or decayed 4-28-66: unannounced payload
Molniya 1C	1966 35A	4-25-66	710	310	24,544	64.5	In orbit: comsat, also transmitting cloud-cover photos
Cosmos 116	1966 36A	4-26-66	92	183	297	48.4	In orbit: unannounced payload

CHRONOLOGY OF SOVIET SPACE FLIGHTS

Name	International Designation	Date	Payload Weight	Period	Perigee	Apogee	Inclination	Status
Cosmos 117	1966 37A	5-6-66		89.5	129	191	65	Re-entered or decayed 5-14-66: unannounced payload
Cosmos 118	1966 38A	5-11-66		97.1	398	398	65	In orbit: unannounced payload
Cosmos 119	1966 43A	5-24-66		99.8	136	811	48.5	In orbit: unannounced payload
Cosmos 120	1966 50A	6-8-66		89.4	124	186	51.8	Re-entered or decayed 6-16-66: unannounced payload
Cosmos 121	1966 54A	6-17-66		89.9	130	220	72.9	Re-entered or decayed 6-25-66: unannounced payload
Cosmos 122	1966 57A	6-25-66		97.1	388	388	65	In orbit: metsat launch witnessed by General Charles de Gaulle

CHRONOLOGY OF SOVIET SPACE FLIGHTS 313

Proton 3	1966 60A	7-7-66	92.5	118	391	63.5	Re-entered or decayed 9-16-66: heavyweight research satellite
Cosmos 123	1966 61A	7-9-66	92.2	163	329	48.8	Decayed 12-10-66: unannounced payload
Cosmos 124	1966 64A	7-14-66	89.5	129	188	51.8	Re-entered or decayed 7-22-66: unannounced payload
Cosmos 125	1966 67A	7-20-66	89.5	155	155	65	Re-entered or decayed 8-2-66: unannounced payload
Cosmos 126	1966 68A	7-28-66	90	132	223	51.8	Re-entered or decayed 8-6-66: unannounced payload
Cosmos 127	1966 71A	8-8-66	89.2	127	173	51.9	Re-entered or decayed 8-16-66: unannounced payload
Luna 11	1966 78A	8-24-66	178	99	746	27	In lunar orbit: returned scientific data until 10-1-66

CHRONOLOGY OF SOVIET SPACE FLIGHTS

Name	International Designation	Date	Payload Weight	Period	Perigee	Apogee	Inclination	Status
Cosmos 128	1966 79A	8–27–66		90	132	226	65	Re-entered or decayed 9-4-66: unannounced payload
None	1966 88A	9–17–66		94.6	85	510	49.6	Decayed 11-11-66: unannounced launch, payload exploded
Cosmos 129	1966 91A	10–14–66		89.4	126	191	65	Re-entered or decayed 10-21-66: unannounced payload
Cosmos 130	1966 92A	10–20–66		89.8	131	211	65	Re-entered or decayed 10-28-66: unannounced payload
Molniya 1D	1966 93A	10–20–66		713	301	24,668	64.9	In orbit: fourth Russian comsat, in 12-hour orbit

CHRONOLOGY OF SOVIET SPACE FLIGHTS

Name	Designation	Date					Remarks
Luna 12	1966 94A	10-22-66	205	62	1,081		In lunar orbit: photographed moon, returned scientific data
None	1966 101A	11-2-66	94.6	87	470	49.6	Decayed 11-17-66: unannounced launch, payload exploded
Cosmos 131	1966 105A	11-12-66	89.9	127	224	72.9	Re-entered or decayed 11-20-66: unannounced payload
Cosmos 132	1966 106A	11-19-66	89.3	129	174	65	Re-entered or decayed 11-27-66: unannounced payload
Cosmos 133	1966 107A	11-28-66	88.4	112	144	51.9	Decayed 11-30-66: unannounced payload
Cosmos 134	1966 108A	12-4-66	89.6	133	198	65	Re-entered or decayed 12-11-66: unannounced payload
Cosmos 135	1966 112A	12-12-66	93.5	161	411	48.5	In orbit: unannounced payload

CHRONOLOGY OF SOVIET SPACE FLIGHTS

Name	International Designation	Date	Payload Weight	Period	Perigee	Apogee	Inclination	Status
Cosmos 136	1966 115A	12–19–66		89.4	123	190	64.6	Re-entered or decayed 12–27–66: classified payload
Luna 13	1966 116A	12–21–66		Flight time: 79.7 hours				Landed on moon: returned photos, soil-density data
Cosmos 137	1966 117A	12–21–66		104.3	143	1,069	48.8	In orbit: unannounced payload, 34th Cosmos orbited in 1966
Cosmos 138	1967 4A	1–19–67		89.2	120	182	65	Re-entered or decayed 1–27–67: unannounced payload
Cosmos 139	1967 5A	1–25–67		87.5	89	130	50	Re-entered or decayed 1–25–67: FOBS flight test

Cosmos 140	1967 9A	2-7-67	88.5	106	150	51.7	Re-entered or decayed 2-9-67: believed to be Soyus precursor
Cosmos 141	1967 12A	2-8-67	89.8	130	214	72.9	Re-entered or decayed 2-16-67: unannounced payload
Cosmos 142	1967 13A	2-14-67	100.3	133	846	48.4	In orbit: unannounced payload
Cosmos 143	1967 17A	2-27-67	89.5	127	188	65	Re-entered or decayed 3-7-67: unannounced payload
Cosmos 144	1967 18A	2-28-67	96.9	388	388	81.2	In orbit: meteorological satellite, similar to Cosmos 122
Cosmos 145	1967 19A	3-3-67	108.6	137	1,327	48.4	In orbit: unannounced payload
Cosmos 146	1967 21A	3-10-67	89.2	118	193	51.5	Re-entered or decayed 3-21-67: believed to be Soyus precursor

CHRONOLOGY OF SOVIET SPACE FLIGHTS

Name	International Designation	Date	Payload Weight	Period	Perigee	Apogee	Inclination	Status
Cosmos 147	1967 22A	3-13-67		89.5	123	197	65	Re-entered or decayed 3-18-67: unannounced payload
Cosmos 148	1967 23A	3-18-67		91.3	171	271	71	In orbit: unannounced payload
Cosmos 149	1967 24A	3-22-67		89.8	154	185	48.4	Decayed 4-7-67: earth-oriented meteorological research satellite
Cosmos 150	1967 25A	3-22-67		90.1	128	232	65.7	Re-entered or decayed 3-30-67: unannounced payload
Cosmos 151	1967 27A	3-24-67		97.1	391	391	56	In orbit: unannounced payload
Cosmos 152	1967 28A	3-25-67		92.2	176	318	71	In orbit: unannounced payload

CHRONOLOGY OF SOVIET SPACE FLIGHTS

Cosmos 153	1967 30A	4-4-67	89.3	126	181	64.6	Re-entered or decayed 4-12-67: unannounced payload
Cosmos 154	1967 32A	4-8-67	88.5	116	144	51.6	Re-entered or decayed 4-10-67: believed to be manned precursor
Cosmos 155	1967 33A	4-12-67	89.2	126	178	51.8	Re-entered or decayed 4-20-67: unannounced payload
Soyus 1	1967 37A	4-23-67	88.6	125	139	51.7	Re-entered 4-24-67: recovery attempt after 17 orbits, 25.2 hrs., failed due to fouled parachutes; V. Komarov killed
Cosmos 156	1967 39A	4-27-67	97	391	391	81.2	In orbit: metsat, forms operational system with Cosmos 144
Cosmos 157	1967 44A	5-12-67	89.4	126	184	51.3	Re-entered or decayed 5-20-67: unannounced payload

CHRONOLOGY OF SOVIET SPACE FLIGHTS

Name	International Designation	Date	Payload Weight	Period	Perigee	Apogee	Inclination	Status
Cosmos 158	1967 45A	5-15-67		100.7	528	528	74.0	In orbit: unannounced payload
Cosmos 159	1967 46A	5-17-67	1,177		236	37,655	51.8	In orbit: unannounced payload
Cosmos 160	1967 47A	5-17-67		88.4	88	127	49.6	Re-entered or decayed 5-18-67: FOBS flight test
Cosmos 161	1967 49A	5-22-67		89.8	127	213	65.7	Re-entered or decayed 5-30-67: unannounced payload
Molniya 1E	1967 52A	5-25-67		319	286	24,737	64.8	In orbit: fifth Soviet comsat, in 12-hour orbit
Cosmos 162	1967 54A	6-1-67		89.2	125	179	51.8	Re-entered or decayed 6-9-67: unannounced payload

CHRONOLOGY OF SOVIET SPACE FLIGHTS 321

Cosmos 163	1967 56A	6-5-67	93.1	162	383	48.4	In orbit: unannounced payload
Cosmos 164	1967 57A	6-8-67	89.5	126	199	65.7	Re-entered or decayed 6-14-67: unannounced payload
Venus 4	1967 58A	6-12-67	2,438	Heliocentric Orbit			In solar orbit: ejected capsule for 10-18-67 Venus soft landing
Cosmos 165	1967 59A	6-12-67	102.1	131	958	81.9	In orbit: unannounced payload
Cosmos 166	1967 61A	6-16-67	92.9	176	359	48.4	In orbit: carried out solar radiation experiments
Cosmos 167	1967 63A	6-17-67	89.2	125	178	51.8	Re-entered or decayed 6-25-67: probable Venus probe failure
Cosmos 168	1967 67A	7-4-67	89.1	124	167	51.8	Re-entered or decayed 7-12-67: unannounced payload

CHRONOLOGY OF SOVIET SPACE FLIGHTS

Name	International Designation	Date	Payload Weight	Period	Perigee	Apogee	Inclination	Status
Cosmos 169	1967 69A	7-17-67		87.6	89	129	50	Re-entered or decayed 7-17-67: FOBS flight test
Cosmos 170	1967 74A	7-31-67			90	129	50	Re-entered or decayed 7-31-67: FOBS flight test
Cosmos 171	1967 77A	8-8-67			90	137	50	Re-entered or decayed: FOBS flight test
Cosmos 172	1967 78A	8-9-67		89.4	126	187	51.8	Re-entered or decayed: unannounced payload
Cosmos 173	1967 81A	8-24-67		92.3	174	328	71	In orbit: unannounced payload

Cosmos 174 1967 82A	8-31-67		715	311	24,699	64.5	In orbit: believed to be comsat failure
Cosmos 175 1967 85A	9-11-67	92.2	130	240	72.9	Re-entered or decayed 9-19-67: unannounced payload
Cosmos 176 1967 86A	9-12-67	102.5	128	982	81.9	In orbit: unannounced payload
Cosmos 177 1967 88A	9-16-67	89.3	126	181	51.8	Re-entered or decayed 9-24-67: unannounced payload
Cosmos 178 1967 89A	9-19-67		90	127	50	Re-entered or decayed 9-19-67: FOBS flight test
Cosmos 179 1967 91A	9-22-67		90	129	50	Re-entered or decayed 9-22-67: FOBS flight test
Cosmos 180 1967 93A	9-22-67	90.1	132	230	72.9	In orbit: unannounced payload

CHRONOLOGY OF SOVIET SPACE FLIGHTS

Name	International Designation	Date	Payload Weight	Period	Perigee	Apogee	Inclination	Status
Molniya 1F	1967 95A	10-3-67		712	289	24,606	65	In orbit: sixth Soviet comsat, in 12-hour orbit

Selected Bibliography

ANTONOV, S. M. "Photographic Processes Used in the First Photographs of the Other Side of the Moon," *Artificial Earth Satellites*, June, 1962.

BLAGONRAVOV, A. A. "Investigation of the Upper Atmosphere and Outer Space Made in the USSR in 1962," *Committee on Space Research of the International Council of Scientific Unions [COSPAR] Information Bulletin*, No. 16, December, 1963.

———. "Report on Space Research in the USSR," *COSPAR Information Bulletin*, No. 6, September, 1961.

CAIDIN, MARTIN. *Red Star in Space*. New York: Crowell Collier Press, 1963.

CLARKE, ARTHUR C., and the Editors of *Life*. *Man and Space*. New York: Time, Inc., 1964.

CRANKSHAW, EDWARD. *Khrushchev: A Career*. New York: The Viking Press, Inc., 1966.

DEWEY, ANNE PERKINS. *Robert Goddard, Space Pioneer*. Boston: Little, Brown and Company, 1962.

DMYTRYSHYN, BASIL. *USSR: A Concise History*. New York: Charles Scribner's Sons, 1965.

DOBRICHOVSKY, Z. *The "Vostok" Spaceship*. Washington, D.C.: Library of Congress, August 24, 1962.

FLORINSKY, MICHAEL T., ed. *Encyclopedia of Russia and the Soviet Union*. New York: McGraw-Hill Book Company, 1961.

FRUTKIN, ARNOLD J. *International Cooperation in Space*. Englewood Cliffs, N.J.: Prentice-Hall, Inc., 1965.

GAGARIN, YURI. *Road to the Stars*. Moscow: The Foreign Languages Publishing House, 1961.

GAVIN, JAMES M. *War and Peace in the Space Age: A New Approach*. New York: Harper & Row, Publishers, 1958.

GIBNEY, FRANK, ed. *The Penkovsky Papers*. Translated by Peter Deriabin. Garden City, N.Y.: Doubleday & Company, Inc., 1965.

GOUSCHEV, SERGEI, et al. *Russian Science in the 21st Century*. New York: McGraw-Hill Book Company, 1960.

HAHN, WALTER, et. al. *American Strategy for the Nuclear Age*. Garden City, N.Y.: Doubleday & Company, Inc., 1960.

HUZEL, DIETER. *Peenemünde to Canaveral*. Englewood Cliffs, N.J.: Prentice-Hall, Inc., 1962.

IVANOV, ALEKSE. *The First Four Stages*. Moscow: 1968.

KOSMODEMYANSKY, A. *Konstantin Tsiolkovsky*. Moscow: The Foreign Languages Publishing House, 1956.

KRIEGER, F. J. *Behind the Sputniks*. Washington, D.C.: Public Affairs Press, 1958.

————. "Soviet Astronautics: 1957–1962," RAND memo RM3595–PR, April, 1963.

LEY, WILLY. *Rockets, Missiles and Space Travel*. New York: The Viking Press, Inc., 1951.

NEWELL, HOMER E. "NASA Report." National Aeronautics and Space Administration.

SHELDON, CHARLES S. "The Challenge of International Competition," presented at the American Institute of Aeronautics and Astronautics Third Annual Space Flight Meeting, Houston, Texas, November 4–6, 1964.

————. *Review of the Soviet Space Program: With Comparative United States Data*. Report of the Committee on Science and Astronautics, U.S. House of Representatives, Ninetieth Congress, First Session. Washington: U.S. Government Printing Office, 1967.

SHELTON, WILLIAM R. *American Space Exploration: The First Decade*. Boston: Little, Brown and Company, 1967.

SOKOLOVSKII, V. D., ed., and THOMAS WOLFF. *Soviet Military Strategy*. Englewood Cliffs, N.J.: Prentice-Hall, Inc., 1963.

"Space Research Activities in the USSR," *COSPAR Information Bulletin*, No. 21, December, 1964.

SELECTED BIBLIOGRAPHY

TITOV, GHERMAN, and MARTIN CAIDIN. *I Am Eagle*. Indianapolis, Ind.: The Bobbs-Merrill Co., Inc., 1962.
TOKATY, G. A. "Soviet Space Technology," *Space Flight*, 1963.
TRW Space Log. Redondo Beach, California: TRW Systems Group. A quarterly publication of information drawn from data released by the National Aeronautics and Space Administration, Department of Defense, United Nations, Tass, and other official sources.
TSIOLKOVSKY, KONSTANTIN. *The Call of the Cosmos*. Moscow: The Foreign Languages Publishing House, 1963.
U.S. SENATE. *Legal Problems of Space Exploration: A Symposium*. Washington, D.C.: U.S. Government Printing Office, 1961.
―――. *International Cooperation and Organization for Outer Space*. Washington, D.C.: U.S. Government Printing Office, 1965.
―――. *Soviet Space Programs: Organization, Plans, Goals, and International Implications*. Washington, D.C.: U.S. Government Printing Office, 1962.
VON BRAUN, WERNHER and FREDERICK I. ORDWAY III. *History of Rocketry and Space Travel*. New York: Thomas Y. Crowell Co., 1966.
ZAEHRINGER, ALFRED J. *Soviet Space Technology*. New York: Harper & Row, Publishers, 1961.

Index

Aberdeen Proving Ground, Maryland, 35
ABM. *See* anti-ballistic missile system
Ackley, Gardner, 9
Afanasyev, 68
A-4. *See* V-1
AIAA. *See* American Institute of Aeronautics and Astronautics
"air glow," 120, 171
air-lock assembly of *Voskhod*, 130, 172–173, 174–175, 177, 274–275
Akademgorodok, Siberia, 206, 243
Alexeovitch, Eugene, 107–108
Alexeyevich, Yevgeny, 88
Alouette satellite (Canada), 196
American Institute of Aeronautics and Astronautics (AIAA), 54–55
Anatolyevich, Yevgeny, 94
Anderson, Raymond, 255
Andreyev, Yevgeny, 6, 7
animals in space flights, 65ff., 78, 83, 168, 208–209, 259
A-9, 32
Antarctic Treaty of 1959, 257
anti-ballistic missile system (ABM), 239, 252, 253, 255–256
"Anti-Cosmic Defense," 244
anti-missile-missile, 244–245, 252, 254, 262

antiradiation serum, Russian research on, 209, 258–259
Apollo program, 129, 130, 260, 265, 269, 270, 273, 276
Archimedes moon crater, 76
astronaut training compared with training of cosmonauts, 88–89, 109–110
A-10, 32
Atlas ICBM, 5, 57, 64, 125, 246, 247, 248, 254, 260
atomic bomb, 3, 37–38, 41, 246, 254
Aviation and Space Week magazine, 239
aviation medicine. *See* bioastronautics

Baikonur launch complex, 56–57
Bardin, I. P., 48
Baryshev, I., 244
Bassett, Charles, 275
Bauer, Edward, 76
Bauman Technological Institute, 61, 164
Before Man Steps into Space, 74
Belka (dog), 78
Beller, William, 209
Belyayev, Pavel Ivanovich, in Red Air Force, 174; backup for *Voskhod 2*, 174; cosmonaut training, 176–177; flight in *Voskhod 2*, 178–186; landing and rescue, 186

Berker, Karl, 29, 31
Berry, Charles, 167–168
Bespalov, Lev, 85
Beyond the Earth's Atmosphere, 21
Big Red Lie, The, 75
bioastronautics, 104, 111–112, 147, 157–158, 208–210
biosatellites, 64, 65
Blagonravov, Anatoli A., 41, 196
Blankers-Koen, Fanny, 158
Blazing a Trail to the Stars, 72–73, 213
Bly, Nellie, 157
Bochum Observatory, Germany, 6, 7, 137, 145
booster size of rockets, 261–262
Borzenko, Sergei, 271
Braun, Magnus von, 35
Braun, Wernher von, 5, 27ff., 31ff., 38, 40, 50, 53, 54, 59, 75
Brezhnev, Leonid, 8
Budker, Gersh, 37, 206
Budker, Ludmila, 37
Bulganin, Nikolai, 41
Burchett, Wilfred, 104
Bykovsky, Valeri Fyodorovich, chosen for *Vostok 5*, 82, 152; personality, 152; flight in *Vostok 5*, 154ff.; near-rendezvous with *Vostok 6*, 156; descent and landing, 157

Caidin, Martin, 30
California Institute of Technology, 233
Canopus (star), 76, 229
Cape Canaveral (Kennedy), Florida, 2, 23, 34, 40, 50, 53, 56, 57, 58, 75, 93–94, 114, 133, 141, 217, 246, 259
Carpenter, Scott, 129–130, 136, 160, 163–164, 186
Catcher in the Rye, 108
Centaur booster rocket, 229
Central Intelligence Agency, 238

Chafee, Roger B., 269, 275, 276
Chicago *Daily News*, 142
Chicago Tribune, The, 101
"Chief Constructor." *See* S. P. Korolev
"Chief Spacecraft Designer." *See* S. P. Korolev
Chlorella plant, 259
Chudakov, Andrei, 62–63
Chuikov, Vasily I., 252, 254–255
civil defense, 252, 254–255, 258
Clarke, Arthur C., 38
Clausewitz, Karl von, 264
Cobb, Jeraldine, 158
Cohen, Victor, 72
Collins, Jimmy, 86
Colorado Springs tracking station, 145
communications satellites, 206–207, 257; *see also* military communications satellites, Syncom, Telstar, Transit
ComSat. *See* Telstar
Concoction for Marriage, A, 153
Cooper, Gordon, 93, 118, 142, 157, 160
Corriere della Sera, 6
"cosmic accelerator," 201
cosmic rays, 196ff., 204ff.
Cosmonaut Day, 6, 194
cosmonauts, reported deaths of, in training, 6; training, 7–8, 79, 107–112; training compared with training of American astronauts, 88–89, 109–110; life in Tyuratam, 151ff., 164; *see also* specific cosmonauts
Cosmos program, 9–10, 145–146, 191–192, 195, 196, 205, 206ff., 218, 242, 244, 245, 266; *Cosmos 1*, 191; *Cosmos 5*, 207; *Cosmos 20*, 208; *Cosmos 22*, 195; *Cosmos 41*, 206, 208; *Cosmos 47*, 208; *Cosmos 50*, 208, 244; *Cosmos*

INDEX

57, 208, 244; *Cosmos 72*, 208; *Cosmos 81–84*, 208; *Cosmos 96*, 231; *Cosmos 100*, 207–208; *Cosmos 110*, 208–209; *Cosmos 111*, 208; *Cosmos 133* and *140*, 266–267; *Cosmos 186* and *188*, 210–212; see also KY-49, KY-56
Couch, Virgil, 252
Council of Economic Advisers, 9
Crimean Astrophysical Observatory, 220
Cronkite, Walter, 134, 270

Data Dynamics, Inc., California, 194
Davydov, Victor, 74
Debus, Kurt, 34, 35, 59
dehermetization, 130, 172–173
Department of Defense, 54, 238
depressurization, 130, 172–173
Dergunov, Yuri, 87, 88
Dew Line, 245
Diamond, Edward, 272
Divine, David, 239, 252
Dmytryshyn, Basil, 49
docking. See rendezvous and docking in space
Doctor Zhivago, 49
Dolgov, Pyotr, 6, 7, 272
Doppler effect, 70
Dornberger, Walter, 27, 29, 35
Douglas, William, 117
Dubinin, Nikolai Petrovich, 258–259

early-warning radar stations, 245
earth-satellite paths, 243–244
"earth shine," 120
Economics Achievement Exhibit, Moscow, 25
Egorov, V. A., 213
"eight-day satellites," 208, 242; see also Cosmos program
Eisenhower, Dwight D., 4, 44, 48, 54, 143, 263

Ekonomicheskaya Gazeta, 258
Electron 1 and *2*, 195–197, 198, 201ff.
Electron 3 and *4*, 204
Encyclopedia of Astronautics, 28
Ephremov, Aleksandre Ivanovich, 65
Epstein, Julius, 272
Ericke, Kraft, 59
Estes, Howell M., 246, 247
Explorer 1, 199

Feoktistov, Konstantin Petrovich, illness during space flight, 8, 170–171; chosen for *Voskhod 1* flight, 162, 164; in World War II, 164; conducts experiments in flight, 164, 172, 176; personality, 175
First Four Stages, The, 65
Fischer, Wilhelm, 36
Fortikov, I. P., 27
Fortune magazine, 59n., 238, 243
fractional orbital bomb, 212, 245
Free Space, 18
Freedom 7, 93, 116
Freeman, Ted, 275
From the Earth to the Moon, 66
Future of Religions, The, 12

G forces, 69, 97, 98, 111, 170, 275
Gagarin, Valya (Valentina), 86ff., 100–101, 276
Gagarin, Yuri Alekseyevich, one-orbit flight, 6, 7, 266; on moon team, 8, 159, 275–277; visit to Kaluga, 16; description of S. P. Korolev, 59; cosmonaut training, 79, 81ff., 88ff.; in Red Air Force, 81, 86; leadership qualities, 83–84; selected "prime cosmonaut," 84, 91, 94, 113; childhood and education, 84–85;

writes article on K. E. Tsiolkovsky, 85; wife and children, 86, 88, 91, 149; personality, 87ff., 94ff., 100, 104, 108, 175; meets S. P. Korolev, 89; parachute training, 89; describes *Vostok 1*, 89–90, 95–96; preparation for flight, 94ff., 113–114, 129; liftoff and orbit, 96ff., 114, 117; descent and landing, 97–98, 114, 134; lauded after flight, 98–101, 114–115; visits K. E. Tsiolkovsky's grave, 101; comment on A. Shepard's ballistic flight, 103; capsule communicator for *Voskhod 1*, 166; helps A. A. Leonov in training, 174–175; quoted on V. Komorov's death, 275; death in aircraft crash, 276

Gas Dynamics Laboratory (GDL), 24, 25, 27

Gavin, James M., 54–55

Gazenko, Oleg G., 64, 65, 138, 168, 209–210

GDL. *See* Gas Dynamics Laboratory

Gemini program, 125, 129, 130, 160, 161, 169, 172, 173, 265

Germany, in World War II, 3, 29, 30, 31, 34–35; use of rockets in World War II, 3, 30; rocket research scientists at Peenemünde, 4–5, 31, 34–35; rocket research before World War II, 23, 26–27, 28, 29, 35; *see also* V–1, V–2

Ghali, Paul, 142

Gilpatric, Roswell, 251–252

GIRD, Leningrad, 27, 29

Givens, Edward, 276

Glenn, John H., 11, 91, 94, 104, 111, 115, 117, 118, 127, 136, 141, 154, 160, 269, 270

Glusko, Valentin Petrovich, 27, 28, 60

Goddard, Esther, 26

Goddard, Robert H., 23, 26, 27, 28, 36

Goldstone tracking station, 75

Gorbov, Dr., 168

Great Britain, rocket research after World War II, 36–37

Grechko, Andrei A., 250

Grissom, Virgil I. ("Gus"), 91, 117, 127, 160, 269, 270, 275, 276

Groettrup, Helmut, 36

Gubarev, Vladimir, 258

Guggenheim, Daniel, 26

Hart, Mrs. Philip, 158–159

heat shield. *See* specific manned spacecraft

Hemingway, Ernest, 118, 167

Herman, Leon, 9

high altitude nuclear tests, 138, 139, 198, 207

Hitler, Adolf, 31, 246

Holiday Inn, Cape Canaveral, Florida, 93–94

Horstig, Major von, 29

Huntsville, Alabama, 54

Huzel, Dieter, 31, 34–35

ICBM, Soviet, 5, 55, 56, 57ff., 64, 161, 246, 249ff., 252, 253; United States, 50, 57, 242, 247ff., 252

ICIC, 41, 47, 48

Ilyushin, Sergei, 6, 7

Institute of Strategic Studies (ISS), 238, 253–254, 256

interceptor spacecraft, 212, 245

Interdepartmental Commission for the Coordination and Control of Scientific Theoretical Work in the Field of Organization and Accomplishment of Interplanetary Communications of the Astronomical Council of the U.S.S.R. (ICIC), 41, 47, 48

International Geophysical Year, 50, 58

"Interplanetary Communications," 40

INDEX 333

IRBM, 50, 64, 246, 252, 254
ISS. *See* Institute of Strategic Studies
Ivanov, Alekse, 65, 69, 129
Izvestia, 99, 154, 194

JATO rockets, 26
Jet Propulsion Laboratory (JPL), California, 62, 75, 190, 233
Jet Propulsion Research Laboratory, Moscow, 29n.
Jodrell Bank Observatory, England, 8, 80, 140, 145, 219, 221, 225, 231, 234–235
Johnson, Lyndon B., 9, 238, 243, 252
JPL. *See* Jet Propulsion Laboratory, California
Judica-Cordilla brothers, 6, 7
Jupiter IRBM, 50, 64

Kaluga, Russia, 13, 16, 18, 20, 21, 24, 101, 148
Kamanin, Nikolai, 7, 114, 172, 173, 175, 276, 277
Kapustin Yar (KY), Russia, 56, 67, 102, 208, 242, 245
Karnavkhov, Vitily, 206
Karpov, Dr., 168
Kasyan, 68
"Katyusha" rocket artillery, 3, 30, 38
"K. E. Tsiolkovsky and His Theory of Rocket Motors for Interplanetary Travel," 85
Keldysh, Mstistlov V., 171, 193, 219, 234, 235, 245
Kennedy, John F., 103
Kesselring, Albert, 31
Khlebtsevich, Yuri S., 213
Khrushchev, Nikita, 2, 8, 16, 41–42, 43, 53, 55, 60ff., 71, 79, 81, 87, 88, 91, 99–100, 101, 103, 138, 147, 156, 159, 160, 192–193, 219, 241, 243, 249, 252
Khrushchev, Nina, 101
Kibalchic, 24

Komarov, Vladimir Mikhailovich, death, 6, 162, 210, 269ff., 275; selected to command *Voskhod 1*, 161, 162, 174; in Red Air Force, 162–163; cosmonaut training, 163, 174; family, 163; editor at Novosti Press Agency, 164; backup for P. Popovich, 165; *Voskhod 1* liftoff and flight, 166; *Voskhod 1* descent and landing, 169–170; flight in *Soyus 1*, 268ff.; personality, 269; state burial, 275
Komsomolskaya Pravda, 275
Konecci, Eugene, 135, 270
K-102 ICBM engine, 57
K-103 ICBM engine, 53
Konstantinovich, Nikolai, 89, 95, 150
Korneyev, Leonid, 27
Korolev, Sergei Pavlovich, 1, 7, 14, 59–62, 64, 65, 74, 75, 78, 79, 84, 89, 90, 91, 94ff., 99, 100, 112–113, 116, 119, 122, 125–126, 127–128, 130–131, 137, 146, 151ff., 160–161, 165, 169, 172, 173, 177, 213, 266, 275
Korotokov, Capt., 105, 106
Kostin, Alexei, 20
Kosygin, Aleksei N., 8, 245
Krieger, Fermin, 39–40
Krylov, Nikolai I., 43
KY. *See* Kapustin Yar
KY-49 satellites, 208, 245
KY-56 satellites, 245

Laika (dog), 2, 65–66, 67, 68–69, 83
Lapp, Ralph, 241, 263, 264
Laurentyev, Academician, 243
Lawrence, Robert, 275–276
Lebedinsky, Alexandre, 226–227
LEM. *See* lunar excursion module
LeMay, Curtis, 44
Lend-Lease in World War II, 30

Leninsky Put Collective Farm, 134
Leonard, Jack, 271
Leonidov, K., 215-217
Leonov, Alexei Arkhipovich, chosen for first space walk, 130, 173; in Red Air Force, 173-174; selected for cosmonaut program, 174; on newspaper *Neptune*, 174; family 174; difficulties in cosmonaut training, 174-175, 176; helped by Y. A. Gagarin, 174-175; personality, 174-175; training for space walk, 176, 177; suggestions for modification of *Voskhod 2*, 177, 274; liftoff and flight, 178-186; space walk, 179-184, 187-188, 265, 269; landing and rescue, 186; space paintings, 189
Levantovsky, V. I., 74
Library of Congress, 238
Life magazine, 91, 158, 163
lifting body, 259-260
Lindbergh, Charles A., 26
Lindenberg, Hans, 35
Logachev, Yuri I., 195, 198
London *Daily Express*, 158
London *Evening News*, 267
London *Times*, 54, 239, 252
Louis, Jennifer, 267, 271
Louis, Victor, 267, 270, 276
Lovell, Sir Bernard, 8, 80, 140, 219, 231, 234, 235
Lovett, Robert A., 263-264
lunar excursion module (LEM), 260
Lunar Orbiter moon probe, 232
lunar probes, *Luna 1* (*Mechta 1*), 74ff., 215; *Luna 2*, 76, 215; *Luna 3*, 77, 81, 215, 221; *Luna 4, 5,* and *6*, 221; *Luna 7*, 221, 222; *Luna 8*, 222; *Luna 9*, 222-227, 228, 229, 233; *Luna 10*, 227-228, 232, 233; *Luna 11*, 232; *Luna 12*, 232-233; Lunar Orbiter, 232; Surveyor program, 76, 222, 228-229; Soviet programs, 8, 10, 74ff., 176-177, 213ff., 220ff., 265, 267, 269; moon photographed by Russians, 6, 77, 221; United States programs, 75-76, 265-266; *Zond 3*, 221

McDonnell Aircraft Co., 115, 125
McNamara, Robert S., 239, 245, 252, 255-256
Malan, Lloyd, 75
Malinovsky, Rodion Y., 42-43, 44
Manned Orbital Laboratory Project, 129, 251, 257, 276
manned space flights. *See* Apollo program, Gemini program, Mercury program, Voskhod program, Vostok program
Manned Spacecraft Center, Houston, Texas, 190, 273
Mars probes, 217, 230, 262; *Mariner 2*, 219, 231; *Mariner 4*, 230; *Mars 1* and *2*, 79, 219, 220; *Zond 2*, 230
Martin Company, 50
Marty, Marty E., 12
Mechta 1 (*Luna 1*), 74
Melnikov, N., 81-83, 156
Mercury program, 79, 91, 93, 103, 114, 115, 125ff., 128ff., 131, 132, 160, 161, 169
Meshchersky, Ivan, 24
meteorite hazard in space flight, 130, 167, 171, 228
meteorological satellites, 243
Method of Reaching Extreme Altitude, A, 26
Miami *News*, 101
Microlock, 62
micrometeorite shower, 228
MIG aircraft, 106, 276
Mikhailov, A. A., 48
Mikoyan, Anastos Ivanovich, 49

military communications satellites, 257–258
military potential of space programs, 10, 11, 71ff., 139, 144, 199, 238–264; *see also* antiballistic missile system, antimissile-missile system, fractional orbital bomb, ICBM, IRBM, lifting body, meteorological satellites, military communications satellites, nuclear rocket, orbital bomb, "overkill" nuclear bombs, reconnaissance satellites
Military Strategy, 42, 43
Minneapolis *Tribune*, 72
Minuteman ICBM, 247
missile strength, of Soviet Union, 238, 248, 252, 253; of United States, 247, 252, 253
missile submarines, 245; *see also* Polaris
missiles. *See* anti-ballistic missile system, anti-missile-missile system, Atlas, ICBM, IRBM, Jupiter, Minuteman, Nike-X, nuclear rocket, Polaris, Redstone, Saturn, Scragg, Sprint, Thor, Titan, Viking
Molchanov, Professor, 163
Molniya 1 (*Lightning 1*), 206, 207
moon probes. *See* lunar probes
Morokhov, Igor, 261
MOSGIRD, Moscow, 27, 28, 62
"Mother Cannon," 30
MR-3. *See* Redstone
MRV, 260
M2FL, 259
M-220 computer, 243

NASA. *See* National Aeronautics and Space Administration
National Aeronautics and Space Administration (NASA), 34, 75, 148, 158, 168, 190, 238, 239, 257, 259, 260, 270
National Observer, 239
Nauka i Tekhnika (*Science and Technology*), 190–191
Nebelwerfer rocket artillery, 30
Nedelin, Mitrofan Ivanovich, 43, 246, 251, 261
Nekrasov, 150
Nell rocket, 26
Neptune magazine, 174
Nesmeyanov, A. N., 40, 47, 48, 145
New York Times, The, 9–10, 12, 62, 192, 239, 255
Newsweek magazine, 239, 272
Nike-X anti-missile system, 253, 256
Nikolayev, Andrian Grigoryevich, flight in *Vostok 3*, 134, 137; in Red Air Force, 135; marriage to V. Tereshkova, 135–136, 149, 152–153; cosmonaut training, 136; birth of daughter, Yelena, 135, 136; personality, 136–138, 141; near-rendezvous with *Vostok 4*, 139, 141, 143, 144; communication in orbit with *Vostok 4*, 141; parachute landing, 143; free-floating in *Vostok 3* cabin, 144; quoted at Y. Gagarin's burial, 276
Nixon, Richard M., 240
NORAD tracking station, 78
North American Air Defense Command Space Detection and Tracking System, 7
Nuclear Energy Center, Novosibirsk, 37
nuclear rocket, 261
nuclear weapons, 245ff.; *see also* fractional orbital bomb, missiles, orbital bomb

Oberth, Herman, 23, 26, 27, 36, 38
Ocean of Storms, 223

INDEX

OKB. See Research Design Office
On the Moon, 21–22
09 rocket. See OR–2
"open-skies" policy, 250–251, 253, 256–257
"Operation Backtrack," 37
orbital bomb, 194–195, 212, 245–246, 252–253, 256
orbital inclination, of Sputnik, 62, 70; of Vostok 1, 97 117, 133; of Mercury spacecraft, 133; of Voskhod 2, 178; of Polyet 1, 193; of Electron 1, 203; of Cosmos 100, 207–208, 242; of Soviet reconnaissance satellites, 242, 244; of United States reconnaissance satellites, 244; of Soyus 1, 268–269
Orbiter Project, 50
Orlov, Vladimir, 214–215, 224
ORM–1 liquid fuel engine, 25
ORM–52 rocket, 28
OR–2 rocket, 28
Orwell, George, 263
Osoaviakhim (Society for the Promotion of Defense and Aerochemical Development), 23, 27
Ostromouv, 6, 154
"overkill" nuclear bombs, 253, 254

Parin, N. M., 67, 69, 73
Parin, N. V., 157
Parin, Vasily, 147, 168, 209
parking orbit, 219, 221
Pasternak, Boris, 49, 267
Paulus, Friedrich von, 30
Pavlov, Ivan, 66
Pearl Harbor mentality of the United States, 53, 55
Peenemünde, Germany, 5, 22, 31ff., 34ff., 39, 54
Peenemünde to Canaveral, 31
Pelipeyko, Vladimir, 190–191
Penkovsky, Oleg, 241, 261
Penkovsky papers, 241

Perelman, Yuri I., 27
Peskov, 60
Petrov, Prof., 59
Petrovich, G., 24–25, 29, 143
Pobedonostsev, Y. A., 27
Pickering, W. H., 75
Pioneer 1, 75
Pokrovsky, Georgi I., 41, 194
Polaris missile, 50, 253
Polyet 1, 192–194, 206
Polyet 2, 194
Popovich, Marina Lavrentyerna, 142
Popovich, Pavel Romanovich, 82; flight of Vostok 4, 140ff.; near-rendezvous with Vostok 3, 141, 143, 144; personality, 141; youth, 142; wife, Marina, 142; in Red Air Force, 142; landing of Vostok 4, 143; free-floating in cabin, 144
Powers, Francis Gary, 251
Pravda, 60–61, 75, 87–88, 99, 198
Pravda Ukrainy, 59
President's Space Council, 135, 238, 267
"Problem of Flying by Means of Wings, The," 20
Proton project, 257, 261; Proton 1, 204–205, Proton 2 and 3, 204
Proxmire, William, 256
Purdy, Anthony, 104
Pushkin, Alexander, 81, 103, 106, 118

Quarks, 204–205
Queen Elizabeth II of England, 158
radiation belts. See Van Allen radiation belts
Radiation Genetics Laboratory of the Institute of Biophysics, Moscow, 258
Radio Moscow, 56, 62, 75, 91, 98, 127, 271
RAF. See Royal Air Force

INDEX 337

Ramsey, Norman, 205
Rand Corporation, 39, 238
Ranger program, 222, 227
RD-1, 2, 3: rockets, 29
Reactive Scientific Research Institute, Moscow, 29
reconnaissance satellites, of Soviet Union, 10, 145-146, 190, 193-194, 207ff., 242ff., 257; of United States, 240-241, 242, 244, 245
Red Air Force, 39, 76, 103
Red Army, 29-30, 35, 39, 41-42, 55, 249
Red Artillery, 30
Red China, 38, 244, 254
Redmilcomsats, 257
Redstone rocket (MR-3), 40, 57, 93
re-entry orbital vehicle. *See* lifting body
rendezvous and docking in space, near-rendezvous of *Vostok 3* and *Vostok 4*, 139, 141, 143, 144; problems of, 145, 269, 273; by United States, 146, 265; near-rendezvous of *Vostok 5* and *Vostok 6*, 156; of *Cosmos 186* and *Cosmos 188*, 210-212
Research Design Office (OKB), 24
"Review of Soviet Space Program," 273
Richter, Henry L., Jr., 62
Road to the Cosmos, The, 213
rocket astronomy, 74, 233
rockets. *See* missiles
Rokossovsky, Konstantin, 35
Romanov, Boris, 134
Roosevelt, Franklin D., 14
Rosenfeld, Albert, 205
Royal Air Force (RAF), 32, 34
RP-1 rocket fuel, 21
"Russians Mean to Win the Space Race, The," 59n.
Russell, Richard, 256

Rynin, N. A., 27, 28

Saksonov, Dr., 168
Salinger, J. D., 108
San Gabriel Valley Amateur Radio Club, 62
Sara Shagan missile base, 256
SAS. *See* Soviet Academy of Sciences
Saturn rocket program, 31, 57, 261; *Saturn 5*, 261, 266
Schirra, Wally, 104, 115, 141, 160, 272
Schriever, Bernard, 10, 44, 59, 246, 260
Scientific Review Magazine, 20-21
Scragg rocket, 253
Sea Lab 2, 129-130
Seaborg, Glenn, 205
SECAM communications satellite system, 206-207
Sedov, L. I., 41, 47
See, Eliot, 275
Seryogin, Vladimir S., 276
Severoy, Andrei, 220
Shapiro, Henry, 267
"Sharik," 131, 133
Sheldon, Charles, 267, 273
Shelton, William, 59n.
Shepard, Alan, 79, 91, 93, 94, 100, 104, 116, 117, 127, 160
Shoemaker, Eugene, 233
Slayton, Donald, 163
Smithsonian Institution, 26, 75
Society for the Promotion of Defense and Aerochemical Development (*Osoaviakhim*), 23, 27
Society of Physics, St. Petersburg, 18
Society of Space Travel (VER), 26-27
Sokolovsky, V. D., 42, 43
solar storms, 115
Souvanna Phouma, 192
Soviet Academy of Medical Sciences, 67
Soviet Academy of Sciences

(SAS), 24, 40, 47, 48, 60, 62, 79, 193, 207, 209, 219, 258
Soviet Union, rocketry in World War II, 3, 29–30; after World War II, 3; effect of space achievements, 3ff., 13–14, 103; failures in space, 8–9, 78–79, 89, 217, 218–219, 220ff., 230, 233–234; cost of space programs, 9; education in, 10; scientific institutions, 10; rocket research, 20ff., 25, 27ff., 36, 37, 40ff.; military position after World War II, 38; after Stalin's death, 49; interest in exploration of space, 71–73, 167ff.; women in space, 147; missile strength, 238; *see also* specific manned and unmanned space programs
Soyus 1 (*Union 1*) flight, 210, 266, 267ff.
Space Detection and Tracking System, North American Air Defense Command, 7
"space garbage," 187
Space Materialography, 199
space programs. *See* manned space flights, unmanned space flights, and specific probes and programs
space walk, 172, 175ff., 179–184, 187–188, 265
Spacecraft programs, 78, 79
Spacecraft 3. See Sputnik 6
Sprint anti-missile-missile, 253
Sputnik program, *Sputnik 1*, 1–2, 3, 4, 10, 11, 46ff., 52–63, 64–65, 70, 87, 106, 210, 240, 246; *Sputnik 2*, 2, 53, 65–69, 71, 87; *Sputnik 3*, 5, 53, 71, 87, 200; *Sputnik 4*, 77–78; *Sputnik 5*, 78; *Sputnik 6*, 78–79; *Sputnik 9*, 79; reactions to Sputniks in United States, 52–55, 65; effect of flights in Soviet Union, 55

spy-in-the-sky satellites. *See* reconnaissance satellites; military reconnaissance satellites
Stalin, Josef, 16, 23, 38–39, 41, 60
"Stalin organs" rocket artillery, 30
Stanyukovich, K., 47–48
Stapp, John Paul, 112
Starlite Motel, Cape Canaveral, Florida, 51
"Star Town." *See* Tyuratam
Starfish, U.S.S., 207
State Astronomy Institute, 74
Stengel, Casey, 51
Stevens, Edmund, 267
Strategic Air Command, 249, 252
Strategic Rocket Force of U.S.S.R., 41, 43, 246, 249, 250
Strelka (dog), 78
Stuart, Homer J., 75
Sudetz, Vladimir, 41
Supreme Soviet of the U.S.S.R., 148
Surveyor moon probe, 76, 222, 228–229
SV5D, 260
Syncom communications satellite, 206

Takhtarova, Anya, 98
Talalikhin, Victor, 162–163
Tass, 99, 193, 268, 269, 271
Tat, Die, 158
Technology Week magazine, 239
television cameras, in *Vostok*, 132, 137, 156; in *Voskhod*, 172, 176, 181–182; on *Cosmos 186*, 211; on *Luna 9*, 225
Telstar, 206
Tereshkova, Valentina Vladimirovna (Valya), chosen for space flight training, 8, 148; flight in *Vostok 6*, 125, 155ff.; cosmonaut training, 135–136,

149ff.; marriage to A. G. Nikolayev, 135–136, 149, 152–153, 159; birth of daughter, Yelena, 135, 136, 159; parachute jumps, 148, 149; personality, 152ff., 157; near-rendezvous with *Vostok 5*, 156; descent and landing, 157; propaganda value of flight, 158; on moon team, 159; at Y. Gagarin's burial, 276
Tessman, Bernhard, 35
"There Is a Defense Against the Nuclear Weapon," 254–255
Thompson, Milton, 259
Thor IRBM, 50, 64, 259
Thurmond, Strom, 256
Tikhonrarov, M. K., 27, 28, 40, 60
Tillich, Paul, 12
Time magazine, 93, 239, 247, 252, 267, 271
Titov, Gherman, illness in orbital flight, 8, 117–118, 121, 138, 169; quoted after flight, 12, 214; cosmonaut training, 79, 81, 107–112; backup for Y. A. Gagarin, 92, 94; in Red Air Force, 103ff.; personality, 103–106, 108ff., 115; marriage, 106; chosen for space flight, 107, 112; familiarization with *Vostok*, 113; death of son, Igor, 113; preparation for *Vostok 2* flight, 115–117, 136; liftoff, 117; description of flight, 118–121; manual control of *Vostok 2*, 120; sleeps during flight, 121–122; descent, 122–123; parachute landing, 123; suggestions for modification of *Vostok*, 127–128, 177
Titov, Tamara, 105ff., 113, 120
Titan ICBM, 57, 247, 254, 260
Tolbubko, V. P., 252–253

Toynbee, Arnold, 262
tracking of satellites, 62, 70, 74, 75, 78, 80, 137, 138, 140, 145, 193, 219, 221ff., 231, 234–235, 268
"transfer maneuver," 143
Transit communications satellite, 257
Tsander, Friedrich, 24, 27, 28
Tsiolkovsky, Konstantin Edwardovich, "father of rocketry," 12–13, 14, 38; 1933 prediction of space travel, 14–15; childhood and education, 16–18; diary *Free Space*, 18; inventions, 19–20; rocket research, 20ff., 25, 27, 36, 67; home, 20; science fiction works, 21–22; later years, 23–24; stamp in his honor issued, 48; meets S. P. Korolev, 62; museum in his honor in Kaluga, 101
TV–3, 2
TV–Zero, 58
TY. *See* Tyuratam
TY–51 satellites, 208, 242
TY–65 satellites, 208, 242, 244
Tyuratam (TY) cosmodrome ("Star City"), 56–57, 69, 76, 81ff., 92, 103, 114, 115, 131, 135, 137, 138–139, 149, 152–153, 166, 186, 192, 242, 256

Ugolyok (dog), 208–209, 259
United Nations, 81, 251, 253
United Press International (UPI), 91, 210, 267, 271, 272
United States, in World War II, 3; atomic bomb, 3; myth of Russian inferiority, 4–9, 46, 75, 109, 271ff.; reaction to Russian space achievements, 1, 10–11, 52–55, 62, 70, 71–72, 103, 144–145; rocket research, 23, 26, 37, 40; knowledge of Russian rocket research, 45ff.; Pearl Harbor mentality, 53, 55;

high-altitude nuclear tests, 138, 139, 198; *see also* specific manned and unmanned space programs
United States Air Force, 144, 245, 251, 253, 259, 260, 261, 272
United States Air Force Institute of Technology, Wright-Patterson Air Force Base, Ohio, 163
United States Air Force Systems Command, Inglewood, California, 246, 247
United States Navy, 50
U. S. News & World Report magazine, 239
United States State Department, 163
Unmanned space programs, of Soviet Union, 1ff., 9–10, 11, 46ff., 52–63, 74ff., 87, 106, 146ff., 190–237; of United States, 160, 190, 199, 206; of Canada, 196; of France, 207; of West Germany, 207; *see also* communications satellites, Cosmos program, lunar probes, Mars probes, meteorological satellites, Proton project, reconnaissance satellites, Sputnik program, Vanguard project, Venus probes
UPI. *See* United Press International
Urey, Harold, 72, 226
U–2 incident, 212, 240, 245, 251

Van Allen radiation belt, 71, 196, 198ff., 208–209
Vandenberg Air Force Base, California, 244
Van Dyke, Vernon, 53
Vanguard project, 2, 50–51, 53, 58, 64
Vasilyev, Pavel, 157
Venus probes, 217, 262; *Venus 1*, 79–80, 218–219; *Venus 2*, 221, 230; *Venus 3*, 221–222, 230, 231; *Cosmos 96*, 231; *Venus 4*, 234–236, 237; *Mariner 5*, 234, 236
VER. *See* Society of Space Travel
Verne, Jules, 20, 66, 67
Vernor, Serge Nikolaevich, 65
Vershinin, K. A., 41
Veterok (dog), 208–209, 259
Viet Nam, 253, 254
Viking rocket, 58
Volkhart, Kurt, 26
Voskhod program, life-support system, 130, 161–162, 172; air-lock system for space walk, 130, 172–173, 174–175, 265, 274; design, 160ff., 171, 172, 273; temperature-regulating system, 161–162; orbital-flight orientation system, 162; compared with Mercury and Gemini programs, 161, 162, 169, 172, 173; cabin instruments, 162; shirtsleeve environment, 166, 171, 173; scientific experiments and observations, 167–169, 171, 172, 187ff.; heat shield, 170; television system, 172, 176, 181–182; Russian people's reaction to flights, 172; development of autonomous navigation system, 176–177; solar orientation system, 185; manually controlled landing, 185–186, 188; as reconnaissance satellite, 242, 257
Voskhod 1, 8, 145, 161ff., 166ff., 176, 265, 274; *Voskhod 2*, 7, 145, 146, 173ff., 221, 265, 274
Vostok program, automatic control system, 78; tests of spacecraft, 93; comparison with United States spaceships, 124ff., 128–129, 130ff.; cosmonauts' suggestions for

modification, 127–128; atmosphere constituents, 129–131; temperature regulating system, 131; optical alignment device, 126, 131; flights, 131ff.; communications system, 132; television system, 132, 137; life-support system, 116, 129–131, 132, 133; manual control of, 132, 133; impact-point device, 132; reentry and landing, 132–134; orbital inclination, 133; heat shield, 98, 125–126, 133; compared with Voskhod, 161; as reconnaissance satellites, 208, 242, 244
Vostok 1, 79, 89–90, 92, 95–96, 97–98, 110, 114, 124; *Vostok 2*, 115ff.; *Vostok 3*, 134ff., 145, 154; *Vostok 4*, 139ff., 145, 154; *Vostok 5*, 152ff., 193; *Vostok 6*, 125, 145, 151, 193
Voyager space probe, 237
V–1 (A–4) bomb, 31
V–2 bomb, 3, 30, 31–32, 35, 36, 37, 39, 57, 58, 246

walk in space. *See* space walk
Washington Star, 271
"Weapons Culture, The," 264
weather satellites. *See* meteorological satellites
Webb, James E., 219
weightlessness in space, effect on animals, 68ff., 209; effect on humans, 97, 109, 117, 121, 122, 138, 144, 157, 169, 176, 177, 180ff., 188, 209–210
Weisman, Walter, 40
Wells, H. G., 66
Welsh, Edward C., 212
Western Test Range, California, 239

White, Ed, 130, 188, 269, 270, 275, 276
Williams, Clifton, 275
wingless wedge. *See* lifting body
With a Rocket to the Moon, 74, 145
Wolff, Waldermar, 36
women in Soviet society, 147ff.
World Peace Council, 1953, 40, 47
World War II, 3, 29ff., 34, 55; *see also* Germany, Soviet Union
Wurfgerat rocket artillery, 30

Yardley, John, 115, 125–126
Yazdovski, Vladimir, 64, 65
Yegorov, Boris Borisovich, illness in orbital flight, 8; chosen for *Voskhod 1* flight, 164–165; as physician, 142, 165; cosmonaut training, 165, 176; family, 165; scientific experiments conducted during flight, 167–169; research in antiradiation serum, 209
Yemelyanov, Dr., 168
Young, John, 160
Yuganov, 68
Yzer optical device, 126, 131

Zabelin, Igor, 12
Zeiber, Albert, 35, 36
Zeppelin, Count Ferdinand von, 19
Zhukov, Grigory K., 44
Zhukovsky Air Force Engineering Academy, 163, 174, 194
Zhukovsky Red Air Force Base, 113
Zond 2, 230
Zond 3, 221
Zvesdochka, 83, 91
Zygielbaum, Joseph L., 75, 194–195, 266, 271